DESIGN STUDIO
2021 VOLUME 2

Design Studio Vol. 2 Intelligent Control: Disruptive Technologies is printed on Fenner Colorset Nero 270gsm and Revive Offset Recycled 120gsm paper, FSC® Recycled 100% post-consumer waste. Papers Carbon Balanced by the World Land Trust. Printed with eco-friendly high-quality vegetable-based inks by Pureprint Group, the world's first carbon neutral printer.

© RIBA Publishing, 2021

Published by RIBA Publishing, 66 Portland Place, London, W1B 1AD

ISBN 9781859469705
ISSN: 2634-4653

British Library Cataloguing-in-Publication Data
A catalogue record for this book is available from the British Library.

Commissioning Editor: Alex White
Assistant Editors: Clare Holloway and Lizzy Silverton
Production: Sarah-Louise Deazley
Designed and typeset by Linda Byrne
Printed and bound by Pureprint Group Ltd
Cover image courtesy of The Living

www.ribapublishing.com

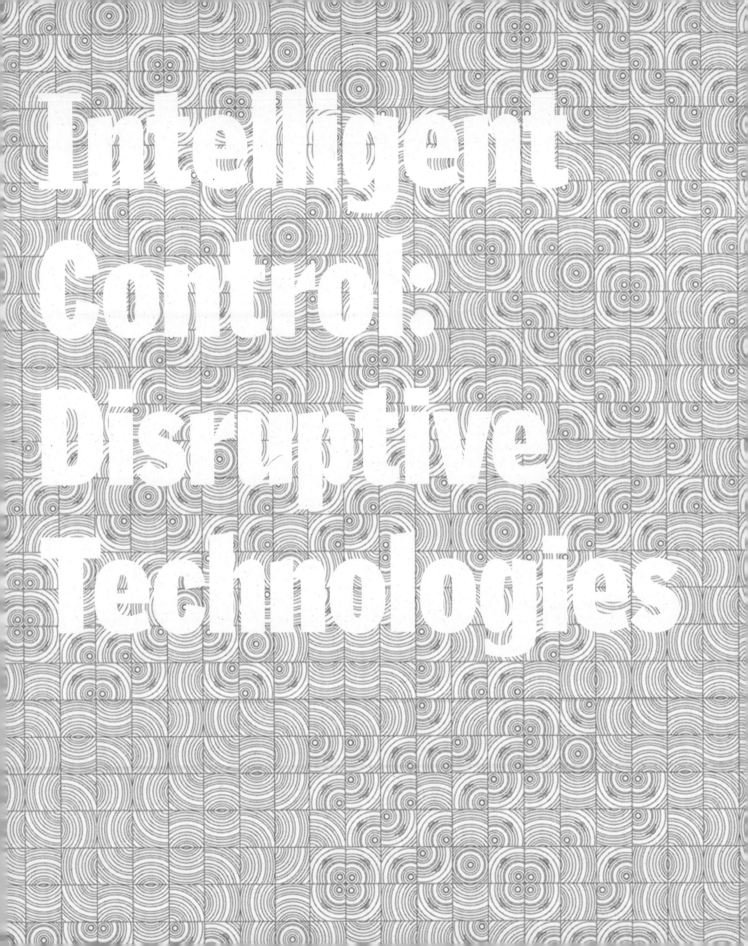

Intelligent Control: Disruptive Technologies

About the Editors

Rob Hyde is a chartered architect and academic at the Manchester School of Architecture, where he co-founded/co-directs the Complexity, Planning and Urbanism research laboratory [CPU]lab and co-leads its aligned taught design studio [CPU]ai. Previously leading the master's Professional Studies programme as well as on Employment, Employability & Enterprise and on Internationalisation, his current leadership is on Knowledge Exchange and impact in relation to applied research + innovation and enterprise.

Operating at the convergence of academia and industry, his research interests focus on Socio-Technical Transitions and Urban Transformations into Sustainable Future Cities/Place and the stakeholders, organisations and transdisciplinary Professional Identities/knowledge that shape it. He is professionally active across diverse cross-disciplinary networks/committees including GMCC, AIA and RIBA. A member of RIBANW Regional Council [Chair Practice & Education Committee] and RIBA Education Committee, he is currently RIBA Research + Knowledge Champion for the Presidents Fact-Finding Mission forming the foundation to RIBA's 2034 masterplan strategies.

Recent publications include *Defining Contemporary Professionalism* (RIBA Publishing, 2019).

@RobHydeRIBA
twitter.com/RobHydeRIBA
instagram.com/RobHydeRIBA
linkedin.com/in/RobHydeRIBA

Filippos Filippidis is an architect and computational design specialist working at the intersection of design and technology. His experience in architectural practice has varied in design scales ranging from web and desktop applications to material research and installations to residential buildings and infrastructure projects. He has worked at firms such as ecoLogicStudio, RoboFold, Foster + Partners and Bryden Wood, and has taught design studios and workshops at the University of Brighton, the Manchester School of Architecture and the AA Visiting School in Melbourne. He is the founder of Continuous-Inputs, a design technology consultancy, and is an Associate Lecturer at the Manchester School of Architecture, where he teaches the MArch [CPU]ai atelier.

@ phil_pidis
twitter.com/phil_pidis
instagram.com/phil_pidis
linkedin.com/in/filipposfilippidis

Acknowledgements

We wish to acknowledge our indebtedness to Aneta and Danae for their ongoing encouragement, particularly over this last year. Our thanks also to Helen Castle, Alex White, Clare Holloway and Sarah-Louise Deazley, along with all at RIBA Publishing, for their constant professionalism, our families for their continued support and, of course, to our fantastic collaborators, without whose inspirational work this book would not have been possible.

We aim for students to enter practice in a state of 'Intelligent Control' in the specialisms they engage with to disrupt current practices and help drive the constant repositioning of the architect.

Mahmud Tantoush, Master's design project with the atelier CPU (Complexity, Planning and Urbanism), Manchester School of Architecture, 2018. Ecologically resilient architecture - generative structural solutions for interacting adaptive elements and flexible programmes. Designed using Grasshopper 3D and Python programming.

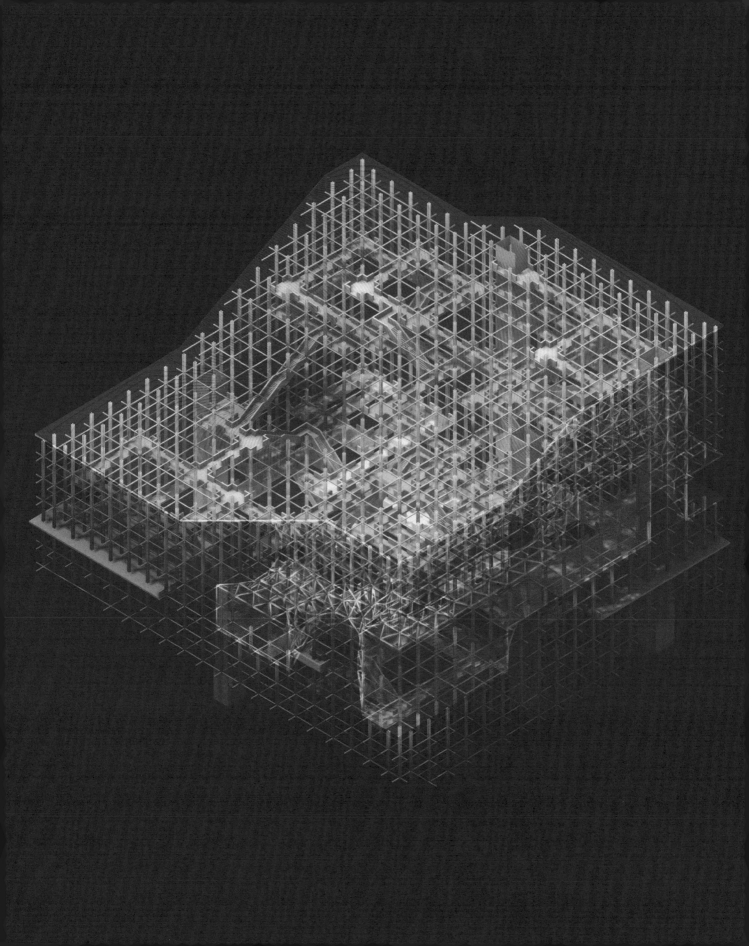

Editor's Note

Rob Hyde and Filippos Filippidis

As a society, we are in a paradigm shift from the digital disruption of the Third Industrial Revolution as it evolves into the automation and smart technologies of the Fourth Industrial Revolution.[1] In this context of 'disruptive innovation',[2] we are bombarded with an ever-expanding and bewildering lexicon of new buzzwords. Each development simultaneously brings threats created and opportunities afforded by rapidly emerging technological advancements in everything from day-to-day working practices within both academia and practice to the great existential challenges faced by humanity.

The Architecture, Engineering and Construction (AEC) industry is increasingly influenced and affected by these wider societal shifts and associated debates, but is historically extremely slow to change its 'lagging performance ... a direct result of the fundamental rules and characteristics of the construction market and the industry dynamics that occur in response to them'.[3] This is exacerbated by a difficulty in harnessing opportunities afforded by disruptive technologies that could tackle long-standing systemic issues within the industry, such as productivity,[4] in order to demonstrate value through both efficiency and effectiveness within or beyond current areas of operation.

As part of this wider AEC ecosystem, architects find themselves restricted by its slowness, which is compounded by their diminishing influence within it. However, technological disruption provides opportunity for the architect to evolve and address current marginalisation. This can be achieved by demonstrating value in both traditional and non-traditional roles, services and diverse fields of operation through enhanced and increased productivity, creativity, design potential, evidence-based decision-making, streamlined delivery, collaboration, stakeholder engagement and much more. Repositioning the role and definition of the architect in both how and where we practise could also help drive the advancement of the wider AEC sector by addressing key internal systemic issues.

A digitally fractured landscape

The pervasiveness of disruptive technologies has given rise to increased speculation about what it means – and will mean – to be an architect. Where and how should we operate in academia and practice? How should we train? What should we learn? Who are our clients? What is our value?

Such questioning is only natural, since architects sit within an ever-evolving field of expanding, overlapping and dissolving professional boundaries. Both the role

Procedural Liveability. Screenshot from game by Aafreen Fathima, Videogame Urbanism 2020–21, a game exploring resilience and liveability of our cities and their ability to adapt to adverse conditions.

Rather than seeing technological disruption as a threat, it should be embraced for the transformative opportunities it allows.

Above: *Exaster. Orientation aware aggregation studies.* From a series of morphological explorations by Marios Tsiliakos, 2020–21.

Opposite: *AID – Architecture for Infectious Diseases.* A generative design exploration catalogue by Connor Forecast, Max Jizhe Han and Jack Seymour at [CPU]ai Atelier Manchester School of Architecture 2019–20.

and definition are constantly changing in response to disruption, be it political, economic, social, technological, legislative or environmental. Historically, we have witnessed many technological advancements that have allowed for great innovations in the built environment. One such example is the Industrial Revolution, which introduced new materials such as cast iron, glass and steel. These radically changed the size, form and function of structures, as well as systems and processes, reshaping the architectural landscape and repositioning the role of the architect. New technologies were adopted rapidly as they presented major opportunities that would benefit and reflect a global growing economy. Rather than seeing technological disruption as a threat, it should be embraced for the transformative opportunities it allows.

Architecture has long explored disruptive technologies in academia and practice, which can be seen in an ever-evolving digital tool-set.[5] It spans decades, from the early beginnings of Computer Aided Design to Algorithmic and Parametric Design, Modelling and Simulation, Digital Fabrication, Virtual and Augmented Reality and currently the shift towards Artificial Intelligence with Machine Learning and Big Data. However, the landscape is fractured and disconnected with technologies being absorbed, adopted and utilised at varying levels and speeds, with little of the potential value afforded being meaningfully realised.

While there are those across academia and practice who have engaged and enabled work with new tools and technologies – such as designing algorithms, robotic systems or new materials – there are many others with a 'fit for purpose' narrow mindset, which results in missed opportunities. A prime example of this is the reliance on volumes of drawing packages to deliver building specifications rather than fully parametric digital twins, which would greatly impact productivity (efficiency and effectiveness) in the iterative process of design, methods of procurement and the construction process.

A narrow mindset has understandably been driven by market forces and the increasing complexity of construction projects which exist in a tightly regulated environment where the lowest price rules in tenders. Such an environment restricts opportunity for more expansive mindsets of innovation or risk-taking. However, we are currently witnessing a tipping point of knowledge that has been cultivated over many years through research within and between the avant-garde in academia and practice, as well as digitally enabled graduates working in mainstream practices. Such a mindset comes with a shift in understanding that architects should be able to engage beyond traditional domain knowledge and with the true complexity of their projects, understanding their interventions and contexts as digital and physical

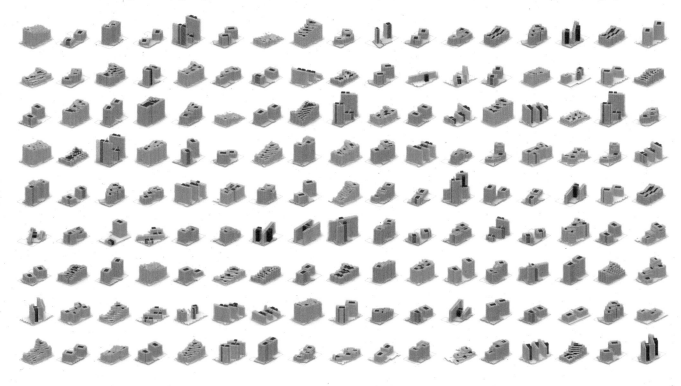

overlapping infrastructures of hard and soft systems with processes interacting and changing over time.

Better informed architects would be of benefit to both clients and the wider society in identifying and understanding underlying problems, latent opportunities and the potential unforeseen impact of proposals on the built environment and beyond. Opportunities include further utilisation of data for new insights, demonstrating the value of the architect at a much earlier or later stage of the development process and connecting these together for informed decision-making. Or for developing better platforms to afford new forms of agency, extended stakeholder participation and wider collaboration, allowing better communication and new knowledge creation between stakeholders, including users, clients, authorities, consultants, professions, disciplines and industries.

In our own teaching, we acknowledge the importance of this mindset. Graduates will likely work for clients who do not yet exist, in roles that do not yet exist, using tools that do not yet exist. We reject the myth of the 'oven-ready graduate'. Instead, we create a culture of exploration and experimentation to develop specialisms beyond the discipline in parallel to the core competencies of the discipline. For future resilience, students are encouraged to develop their skill sets across diverse areas of knowledge rather than in a single disciplinary silo, to have adaptability to alternative future contexts and anticipate technological (and other) disruption. Ultimately, we aim for students to enter practice in a state of 'Intelligent Control' in the specialisms they engage with to disrupt current practices and help drive the constant repositioning of the architect.

Intelligent Control is a term borrowed from the area of 'Control Systems', which utilises Artificial Intelligence computing techniques. Here, we reinterpret it as a relational mindset that engages in knowledge transfer from a multitude of diverse disciplines, with the ability to exchange and make use of information between different modes, models and mediums. This is driven by the motivation for an increase in the design space which allows for more factors to negotiate earlier in the design process.[6] It is characterised by a bottom-up and data-driven approach for design in which human bias is taken out of the equation and which could shed light on underlying inter-relationships and dynamics. The outcome of these negotiations lead to a process-driven complex result that is both physical and digital, better able to enrich our built environment and the challenges faced.

Agents of change

This volume of the 'Design Studio' series showcases the architect as a key agent of change, harnessing technology and enabling disruption both within and beyond the traditional confines of our profession and industry.

The assembled contributors represent a broad group of academics and practitioners with the mindset to challenge norms with their continued experimentation. Their work embodies Intelligent Control within their emergent specialisms, which are disrupting and have the potential to further disturb current practices and modes of working across areas such as Generative Design and Artificial Intelligence, Design Automation, Simulations for Urban Growth, Gamifying Architecture, Digital Augmentation – AR/VR, Digital Fabrication and Robotics and New Material Practices.

The book is divided into three key sections:
· The first section looks at inspiring design studios and taught courses worldwide that engage in specialist areas of technological disruption.
· The second section focuses on research and development in practice that tests potential application in professional settings.
· The third section contains building case studies which demonstrate application in the delivery and quality control of constructed projects.

Key themes emerge, including the access to digital tools and automation affording greater democratisation and an increased collaborative approach in design. Mollie Claypool explores how access to design and making processes readily available to a wider set of stakeholders enables action on issues around social justice, including localised manufacturing and circular economies. Sandra Youkhana and Luke Caspar Pearson search for unexplored possibilities by leveraging the creation and use of architectural video games to speculate on new forms of urban design through

Right: Computational Design Applied in Practice, [CPU]ai Atelier Manchester School of Architecture. Academic workshop by Karim Anwar, Creative Technologies, Bryden Wood, October 2019.

Left: Robotic construction of interlocking facade panels, Plexal. Integrating advanced workflows into practice. Professional workshop by Eva Magnisali of DataForm Lab, February 2020.

engaging a large number of participants, including non-professional stakeholders. Lassa Architects demonstrates the value of state-of-the-art manufacturing techniques applied to the design and build of small-scale projects.

In terms of advancement of sustainable design futures to potentially address the demands of an ever-increasing population, Complexity, Planning and Urbanism [CPU] acknowledges this as a wicked problem and speculates on possible future trajectories by researching cities as complex adaptive systems. Mette Ramsgaard Thomsen advocates for a rethink of sustainable building practices at a different scale through bio-based material paradigms to push us towards more sustainable futures. The Living investigates climate change through research as practice, by using real-world scale prototypes to test, validate, learn and iterate, refining the process with each cycle. Odico uses advanced fabrication techniques to allow for the manufacture of greater design complexity while enhancing sustainability for the built environment.

In respect to the human designer when interacting with the machine, Soomeen Hahm raises that the human role has been often neglected in the discussion of computational design and automation, and that there needs to be a rethink of the role of humans in the production chain, whilst Danil Nagy emphasises the dialogue and information exchange between human and computer. Like every tool, Danil explains why these new technologies will not replace the designer but will give them new capabilities to improve and optimise their design process.

The importance of Data Driven Design (DDD) to leverage information early on in the design process to inform and reveal opportunities and constraints is demonstrated in the City Intelligence Lab (CIL) use of DDD processes and simulations which it combines with Artificial Intelligence to enable a deeper understanding

of the intertwined environmental, social and spatial effects of our constant urbanisation. SHoP – learning from past projects – research initiatives and innovation in other disciplines, and integrate management, visualisation and tracking of data to ensure the success of the project from concept to construction.

Hawkins\Brown shows us the positive effects of a practice's capacity to absorb and apply by taking us through an exemplary project serving as a catalyst for new knowledge accumulation that would alter the practice's internal structures and processes through cultural change.

Finally, Stefana Parascho and Bryden Wood advocate for a need to get out of our comfort zone, to venture into neighbouring disciplines and bring back lessons learnt. This will allow for a greater bandwidth to push our discipline's boundaries and bring potential greater value.

Reflection

The contributors included represent a small sample of those with a mindset to engage with disruptive technologies and demonstrate a level of Intelligent Control. By the time this book is published, we expect that many of these areas will have developed further or evolved into something else entirely. However, the critical importance of this volume is that it provides contemporary examples which we hope will inspire the reader to further explore how these specialisms have emerged; to track them forward into the future, to identify the enablers and barriers to their development, to recognise opportunities and their value, and to provide encouragement to engage with disruptive technologies and emergent specialisms to uncover and address both current and future challenges.

Ultimately, our hope is the reader will feel empowered to reflect on their own personal trajectories and the possible future contexts in which they will operate, where to concentrate their efforts for maximised impact and to consider the future possible position/s and definition/s of the architect.

1. Schwab, K., *The Fourth Industrial Revolution*, Crown Publishing Group, New York, 2017.
2. Bower, J. and Christensen, C.M., 'Disruptive technologies: catching the wave', *Harvard Business Review*, vol. 73, issue. 1, Jan/Feb 1995, pp. 43–53.
3. Ribeirinho, M.J., Mischke, J., Strube, G., Sjödin, E., Blanco, J.L., Palter, R., Biörck, J., Rockhill, D. and Andersson, T., 'The next normal in construction. How disruption is reshaping the world's largest ecosystem', *McKinsey & Company – Reports*, 2020, p. 5, accessed 12 January 2021.
4. Barbosa, F., Woetzel, J., Mischke, J., Ribeirinho, M.J., Sridhar, M., Parsons, M., Bertram, N. and Brown, S., 'Reinventing Construction: A Route to Higher Productivity', *McKinsey & Company – Reports*, 2017, accessed 12 January 2021.
5. Claypool, M., 'The Digital in Architecture: Then, Now and in the Future', *SPACE10*, 2020.
6. Pigram, D., Interview at IN(3D)USTRY Conference, 'From Needs To Solutions', 2016.

Ulysses Sengupta,
Eric Cheung,
Solon Solomou,
Sigita Zigure,
Mahmud Tantoush,
May Bassanino
and Rob Hyde

Complex Urban Futures:
Design Science for Flux Territories

The twenty-first century commences with cities as the nexus between an urban age, the Fourth Industrial Revolution and the tipping point for catastrophic climate change. The need for sustainable future cities has never been more critical. The professions and disciplines involved in the design of future cities appear powerless in the face of:

· the complexities involved
· the diminishing role of the designer as an instigator of change in the face of quantifiable economic priorities, and
· the oxymoron of sustainable development.

The default passive position taken by most designers in adherence to, or intellectual manipulation of, existing minimal regulatory frameworks pushes the responsibility for sustainability back towards government. While this is a form of capitulation, the root vulnerabilities based on the difficulties of adopting a critical position that can impact negatively on one's selected vocation are exacerbated by:

· contradictions presented in the idea of sustainable development, and
· the related lack of understanding of the implications of design propositions.

This contested space is illustrated through opposing and parallel interpretations of sustainable development, e.g. as a strategy based on recognising the limitations of growth in the context of unrenewable resources[1] and/or the United Nations' economic perspective setting out a significant number of its Sustainable Development Goals (SDGs) around growth and prosperity.[2]

The promise of anticipated technological solutions for urban sustainability – proliferated through Smart City, Internet of Things (IoT) and Big Data perspectives – continues to encourage a business-as-usual approach to the design of future cities. There is a limited amount of alternative engagement with designed futures through the acknowledgement that more information can lead to more informed decisions, some optimisation-based efficiencies and advances in automation. However, the majority of designers continue to concentrate on aesthetics and immediate functionality rather than engagement with challenges at multiple scales. Even the designers deliberately engaged in attempts to reduce the negative impacts of the built environment on climate change are limited to material and functional specifications aimed at reducing the potential damage caused at a local level.

Setting aside the need to redefine the design professions as ethical vocations that need to contribute towards sustainable future cities for another day, the entry point for designers wishing to contribute remains limited, due to the overwhelming cognitive blocks in confronting urban sustainability in relation to climate change. The lack of engagement stems from deficiencies in the ability to usefully comprehend the implications of design contributions within the complex, multiscale, temporal and emergent phenomena that constitute the contemporary urban process, resulting in a wicked problem that defies clear definition.[3]

A research-based design perspective

Engaging with complex systems

Urban Transformations is an emergent interdisciplinary field of research combining complex systems and urban studies. The methodological developments in this area encourage engagement with new opportunities, advancing desirable transitions towards sustainable futures. While focused on socio-economic research, the emphasis on the consideration of real-world impact of disruptive and projected technologies and urban phenomena provides a useful basis for the development of an alternative 'design system'.

Engagement with complex systems from an urban transformations perspective enables the understanding of the role of interventions within larger temporal systems with existing dynamics. This is as essential for design systems engaging with climate change as it is for influencing trajectories of socio-technical transitions, such as the growth of cyber-physical systems and automation based on urban data collection and analysis.

Complexity, Planning and Urbanism [CPU]

The Complexity, Planning and Urbanism [CPU] group at the Manchester School of Architecture consists of the [CPU]lab and [CPU]ai design master's and undergraduate design studio atelier, with a founding role in the transdisciplinary ESRC research network DACAS (Data and Cities as Complex Adaptive Systems) spanning Japan, China, Brazil and the UK. At the [CPU]lab, urban transformation is researched by combining complexity theories and the development of new digital tools, allowing simulation and experimentation of previously impracticable temporal urban phenomena. As complex systems are an interdisciplinary area of research, concepts from physics, economics, ecology, sociology and computer science are integrated into an evolving body of knowledge aimed at understanding real-world phenomena characterised by temporal change, unpredictability, adaptation and evolution. The wide relevance of the research is demonstrated through the variety of funded research undertaken on ICT-enabled sustainable Smart Cities, Connected Autonomous Vehicle (CAV) futures and the use of IoT data for more agile governance.

The designers and researchers of the [CPU] group have established a design science research approach for positioning urban and architectural interventions within a computationally enhanced systemic formulation of the design problem. This design system purposefully centralises the often ignored or superficial consideration of multi-scale temporal dynamics, disruptive possibilities and relational considerations of the urban process. The urban is a complex adaptive process.[4] Understanding existing trajectories of change and influencing them towards desirable futures through design requires an engagement with the competing, contradictory, non-linear, emergent, self-organising and open-ended phenomena that are fundamental to this. Rather than operating from separated positions for research and design, [CPU] explores the space between design and science by merging the real and the artificial through future studies, data analysis, computational simulation, digital participation and research into cities as complex adaptive systems.

[CPU]ai (the master's and undergraduate design studio atelier) orientates the design studio in architectural education specifically towards learning and formulating new approaches rather than reusing tested methods without modifications on new problems. This builds the capacity for computation-enabled complexity-based design thinking in future designers over a two-year immersive process. Design projects make use of information from multiple sources and knowledge from the sciences to translate theoretical concepts involving complexity, resilience and self-organisation into spatially relevant design strategies. The cognition of systemic urban processes over time – knowable disruptions due to technological transitions and the interconnectedness of complex phenomenon – lead to an understanding of urban processes, where outcomes cannot be directly or fully controlled through intentional design interventions. This forces alternative positions and methods for the introduction of design towards sustainable futures, enabling engagement with otherwise problematic conditions and transformations.

The role of computation

Emerging methods of applied computation play an essential role in the consideration of architectural and urban design interventions situated within dynamic urban systems. The potential of the Fourth Industrial Revolution lies in a designer's ability to utilise computer simulation, gamification, AI, automation, VR and computer-generated design to engage with complex phenomena such as multi-dimensional interactions and feedback cycles between interconnected urban factors. Developing capacity in constructing custom computational models incorporating problem-specific processes requires synthesis of multiple new skills. Attainment of an ability to explicitly model, analyse and dynamically simulate selected urban systems precedes an enhanced awareness of the changes and contradictions in the spatio-temporal processes within the larger systems relevant to any intervention. New interventions are typically introduced as various types of rule-based elements that impact on the process of simulated urban systems which unfold over time. The design system enables the exploration of alternative proposals within simulated urban processes that have identifiable dynamics in play. The construction of such computational thought experiments forces the designer to consider specific relationships and dynamics that play a part in open-ended urban systems, while understanding the strategic possibilities and limitations of their own specific influence on the system.

The design system

The design system can be understood as three relational areas consisting of:
1. problem formulation
2. simulation
3. design resolution.

Each of these areas has multiple internal components with feedback loops, iterations and operational possibilities defined with reference to theory-based concepts from the theoretical framework.

Problem formulation

The design system developed adopts a complex systems perspective as an overarching theoretical framework for all areas. The formulation of a design problem using this lens becomes essential to the design position. The process of problem formulation involves:
· gathering relevant information and data
· problem identification based on context and identifiable issues
· development of an understanding of the problem in the urban context, and
· definition of desirable goals with stakeholders (ranging from government and local authorities to funders and local urban dwellers).

The design system itself is conceived as an amalgamation of three interactive areas: problem formulation, simulation and design resolution, where parts within each relate through iterative feedback loops.

Emerging methods of applied computation play an essential role in the consideration of architectural and urban design interventions situated within dynamic urban systems.

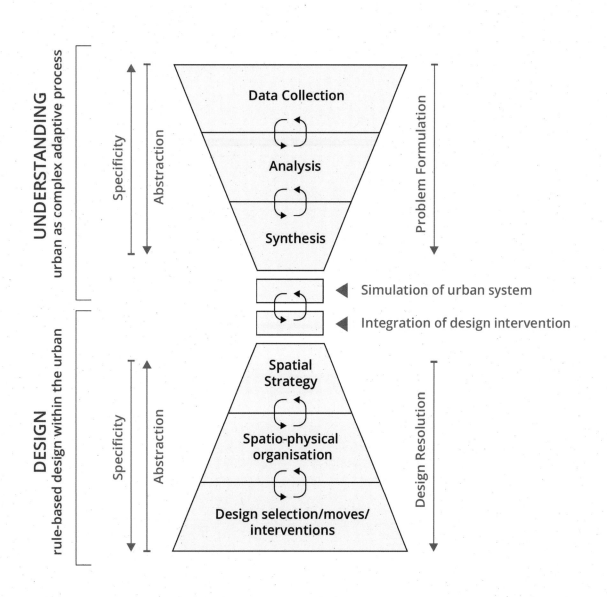

Simulation based on use of Airbnb and Zoopla data to extrapolate housing price changes in Manchester. The figure extrapolates identified correlations of Airbnb and property price growth. By Mahmud Tantoush, Alex Macbeth and Raden Norfiqri.

80% listings airbnb

61% average price increase

2025

Rental property/price   Airbnb listing

The synthesis of multiple potentially competing aspects of change and desire as temporal possibilities situates hybrid socio-physical processes within a structure of systems context consisting of multiple participants, motivations and physical space. Spatial design proposals are aimed at modifying the potential forces of influence and interactions within the simulated complex urban process.

Airbnb Data Analysis: Disruptive technological trends
The project critically investigated property price scenarios in Manchester based on the growth of digital 'sharing economies'. The technological disruption of increasingly bottom-up e-commerce practices was explored through API-based data capture, correlational analysis and temporal geolocated visualisation of Airbnb and Zoopla data. The findings were extrapolated in a simulation to understand projected outcomes until 2050. Findings such as students being priced out of the 'Manchester Corridor'[5] in future scenarios was utilised to formulate the problem and select design strategies aimed at addressing this possibility.

Low Carbon Urbanisation: Walkable activity-based design
The project was based on a multi-criteria analysis with future development positioned within the city and UN (SDG-11) goals aimed at net-zero futures. It spatialised theoretical applications of centrality, density and compactness to enable exploration of highly connected clusters based on walkability and proximity (e.g. to work and amenities). The computational approach generated multiple urban development scenarios in relation to low-carbon urbanisation indicators while problematising the 'known-unknown' of a future residential population demands by developing and using a synthetic population of intelligent computational agents (an applied use of AI research).

Simulated Urban Systems and Simulated Design Interventions
In order to merge design perspectives – the intention to change the status quo or existing trajectories – with models of existing urban systems, two types of computational simulations have to be facilitated:
· The first is based on models enabling computational simulations of the targeted urban systems relevant to the problem formulated.
· The second on rule-based models testing computational simulations of design interventions – that do not yet exist – within the simulated urban systems.
Mental models or mental simulations have been used previously to think about future conditions that are not observable in the world 'out there', based on design

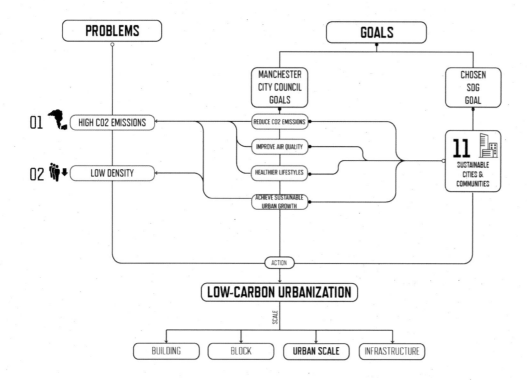

The design problem formulation for low carbon urbanisation identifies common UN (SDG-11) and Manchester City Council goals, before formulating the link to potential actions at the scale of implementation. By Sevdalina Stoyanova and Adrian Dimov.

interventions. However, computational simulation results in an alternative process and provides its own pedagogic and cognitive value. Mental models/simulations incorporate implicit or hidden assumptions, whereas computational models and simulations of urban systems are necessarily explicit. The 'assumptions are laid out in detail, so we can study exactly what they entail'.[6]

Constructing a computational simulation is an intellectual task requiring synthesis of targeted urban systems and potential relationships between these and new design interventions. For a simulation to be executed, the details of relevant processes must be programmed and translated to computer code. This requires consideration of specific relationships and dynamics that might play a part in the behaviour of modelled open-ended urban systems. There is also the need to strategically identify the specific points of interaction that hold the potential for new design interventions to exert influence.

Our perspective on computational simulation consists of algorithmic models of complex urban systems and their dynamic processes and feedback cycles over time. This is distinct from the typical reference to 'simulation' and 'model' in architecture as representations of proposals and the built environment in the form of architectural drawing, digital or physical 3D representational model or visualisation, flythrough and walkthrough.[7] The initial computational simulation of urban systems provides a basis for design-orientated models and simulations aimed at testing alternative rule-based designs and interventions. The notion of simulation from architectural research used here is to examine proposed spatial strategies, organisations and design interventions, 'to "preview" identified future scenarios'.[8]

Strategic Urban Phytoremediation:
Urban ecologies and patch dynamics
The project investigated potential shifts in the future development trajectory of Manchester's Green Quarter, a declining light industrial area. The project aimed to challenge the notion that manufactured and human-centric topologies exclude natural ecologies. A causal loop diagram (balancing and reinforcing loops) was used to position phytoremediation – to address soil contamination – as a temporal ecological factor within existing relational dynamics defined by stakeholders. The introduction of this new factor enabled exploration of processes of transformation using a 'patch dynamics' framework towards instigating higher levels of urban co-existence with natural ecosystems. The computational

approach primarily utilised a cellular automaton approach to simulate the urban system of interest. The simulated design intervention enabled the strategic input of hybrid phytoremediation-based land use and building types, enabling an exploration of dynamically driven alternative scenarios – with readouts on FAR, density and biodiversity potential indicators – over a period of 20 years.

Urban Panarchy: Adaptive cycles for land use towards
reduced energy consumption
The project aimed to estimate and strategically reduce urban energy consumption at a large urban scale. The urban system of interest (based on the Green Quarter, Manchester) was simulated as a constantly responsive adaptive system using a cellular automata approach. This incorporated multiple scales of interaction based on Panarchy, including cycles of land use per plot (e.g. unused, construction, redundancy, dereliction, demolition) and probabilities driving interaction between plots (e.g. proximity of residential development to a landfill site). The simulated design intervention incorporated new ecological phases within cycles of plot development to introduce ecological resilience into the balance of anthropogenic and ecological land uses. An Urban Metabolism approach was used to quantify the impact of temporal interventions and cyclic morphological

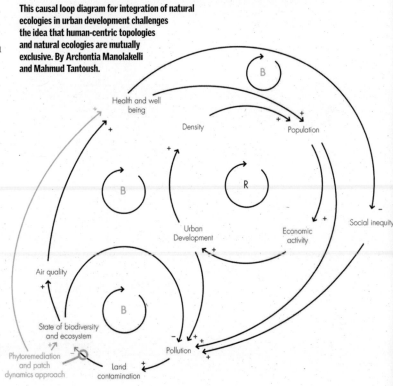

This causal loop diagram for integration of natural ecologies in urban development challenges the idea that human-centric topologies and natural ecologies are mutually exclusive. By Archontia Manolakelli and Mahmud Tantoush.

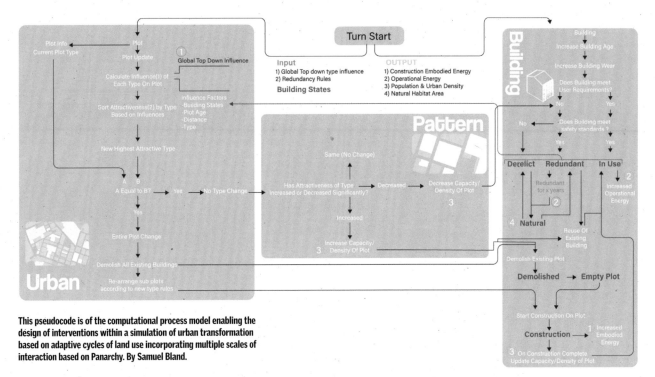

This pseudocode is of the computational process model enabling the design of interventions within a simulation of urban transformation based on adaptive cycles of land use incorporating multiple scales of interaction based on Panarchy. By Samuel Bland.

changes in the dynamic simulation for immediate overall embodied and operational energy use readouts.

Design resolution

The third activity area builds on the knowledge of strategic possibilities for new design interventions within simulated urban systems by formalising potentials for influence compatible with spatial options. The three components involve:

1. The development of abstract spatial strategies into rule-based computational models.
2. The generation of outcomes or spatio-physical organisation from these models using computational methods.
3. The critical selection of appropriate design interventions or strategies based on their effective value towards achievement of design goals.

The three areas within the design system apply variations of gamification, AI (data analysis and intelligent agents) and automation and virtual reality (visual interaction) to enhance both computer simulation and computer-generated design. However, it is the interaction between the problem formulation, simulation and design resolution areas that provides an understanding of design interventions in their potential to generate alternative futures in the context of complex adaptive urban processes.

Centralising public space

The project approaches the problem of sustainability on the basis of neighbourhoods that provide everyday amenities for local residents and hence reduce the need for needless motorised travel. It explores the possibilities of urban development based primarily around access

Single Public Space   **Few Public Spaces**   **Many Public Spaces**

This generative design tool integrated new small-scale public spaces and customised urban patterns to support integrated everyday amenities as the basis of urban morphology, addressing the problem of sustainability through strengthened neighbourhoods. By Lowell Clarke.

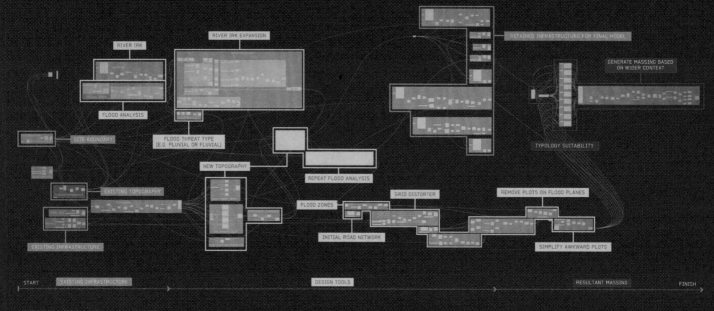

This diagram illustrates the computational process developed to formulate a generative design tool producing alternative flood-resilient spatial strategies. By John Foley.

DESIGN STRATEGY: MARK II.
DESIGN TOOL OVERVIEW.

to small-scale public spaces and related local amenities for the benefit of residents. The generative design tool developed in this study enables the assessment of multiple alternative development scenarios based on different spatial distributions and related urban geometries. It utilises an algorithmic process incorporating urban design considerations to generate desirable urban outcomes driven by the inputs for public spaces. The outcomes are assessed in terms of non-motorised accessibility to public spaces, sensitive to street patterns, urban types and the potential dwellings served.

Flood resilience
The project responded to the statistically rising environmental risk of river flood threats in Manchester, in the context of extreme rain. It undertook a higher resolution analysis than flood risk assessments from government agencies to support spatial design interventions at an urban scale. The generative design tool developed enabled the design of urban areas through an informed manipulation of the topography towards greater flood resilience and control of water run-off. This main driver was part of a process of design automation where the incorporated rules from

known urban development strategies and types were subsequently used to produce alternative future-designed scenarios through the initial manipulation of topography for flood resilience.

MakeMyManchester
The project addressed the difficulties of residents' participation in urban change strategies – especially where these concern long-term societal agendas such as environmental sustainability – by developing a web-based participatory game. The web-based participatory design tool developed using the Unity game engine utilises custom coding of game mechanics and a representation of actual city features to enable players to operate in various modes, such as resident, planner and developer. This gamification approach to understanding the contestations and desires of local populations is complementary to computational simulation and modelling, as it allows the first-hand study of dynamics. It is also a way for designers and resident populations to test their own interventions within a multi-stakeholder and objective (complex) system.

Design and sustainability

Spatio-physical structures generating the conditions that encourage certain human activities is a supplementary area of inquiry into urban environmental sustainability, where the dominant discussion focuses on behaviour change for households in terms of energy consumption and travel activities. Traditional research on behaviour change typically accepts that the environment cannot be adjusted. Architecture and urbanism are in a unique position to engage with sustainable development through reconfiguration of physical arrangements and organisation of space and time in the built environment while integrating the wider considerations of land use-transport integration. The looped influence of spatio-physical structures and behaviour remains a complex problem for sustainable development. By spatialising aspects of social, economic, temporal and technological considerations through abstraction, theory and systemic strategies, it becomes possible to unravel the inherent complexity and contradictions for architecture, urban design and planning processes. The systemic approach enables consideration of both problem formulation and design resolution in terms of relevance and approach across multiple scales ranging from users to policy.

The design system itself is part of evolving methodological research, with current emphasis on the identification of 'better' design strategies and outcomes. The process has particular relevance for governance and strategic planning, due to the creation of multiple design pathways and outcomes that can be analysed in terms of ambition, effectiveness and practicability. It remains distinct from City Science approaches supporting evidence-based decision-making, as it generates future scenarios through design approaches that demonstrate avenues to change existing trajectories of transformation.

The integration of a complex systems perspective with computational thought experiments is of particular importance when designing for sustainable futures that can be influenced by emerging technologies such as cyber-physical and automated systems. Sustainable development deals with open-ended urban systems involving synergies and contradictions. Generation and examination of alternative systemic scenarios allows identification of different merits in different dimensions, e.g. the trade-offs between socio-economic activity and environmental impact.

The unravelling of complex urban processes as temporal and adaptive systems, cognition of the role of design and ability to incorporate or respond to disruptive technologies provides designers with an alternative position from which to engage with climate change and sustainable development. The UK and the EU have committed to becoming climate-neutral by 2050. The framework developed here situates design contributions for a sustainability transition within systems of emerging technology, economics and social equity.

Looking ahead, it is essential for designers to embrace processes involving computation and AI if they are going to engage with societal issues from a position of knowledge and expertise rather than as passive bystanders. The advancement of processes and methods based on additional skills and robust research takes time. Practice and academia need to develop a new space – based on mutual agreement to purposefully evolve the future of design – for design and science to exist together.

Web-based participatory design tool to create cities based on serious role-play game methods. Gameplay is customised to specific possibilities based on different roles, including urban dweller, local authority, developer, etc. By Patrick Lyth and Jordon Lambert.

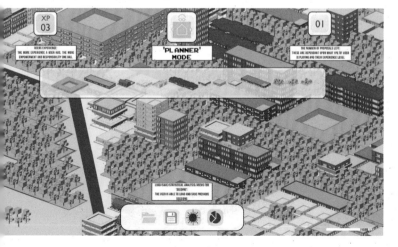

1. Meadows, D., Meadows, D. and Randers, J., Limit to Growth: *The 30-Year Update*, Chelsea Green Publishing, 2004.
2. UN General Assembly, *Transforming our World: The 2030 Agenda for Sustainable Development*, 21 October 2015, A/RES/70/1, available at: https://www.refworld.org/docid/57b6e3e44.html, accessed 8 November 2020.
3. Rittel, H.W. J., and Webber, Melvin, M., 'Dilemmas in a General Theory of Planning', *Policy Sciences 4*, no. 2, 1973, pp. 155–69.
4. Sengupta, U., 'Complexity Science', in Christopher Doll, Deljana Iossifova and Alexandros Gasparatos (eds), *Defining the Urban: Interdisciplinary and Professional Perspectives*, Routledge, London, 2017, pp. 249–65.
5. The Oxford Road Corridor is Manchester's innovation district located south of the city centre. It hosts a concentration of knowledge, business and cultural assets, including the University of Manchester and Manchester Metropolitan University, with a combined population of 70,000 students.
6. Epstein, J.M., 'Why model?', *Journal of Artificial Societies and Social Simulation*, 11.4, 2008, p. 12.
7. See Linda N. Groat and David Wang, *Architectural Research Methods*, John Wiley & Sons, New Jersey, 2013.
8. Iossifova, D., 'Architecture and urban design: Leaving behind the notion of the city' in Christopher Doll, Deljana Iossifova and Alex Gasparatos (eds), *Defining the Urban: Interdisciplinary and Professional Perspectives*, Routledge, London, 2017, pp. 109–27.

The Allegorithmic Utopia of Videogame Urbanism

**Sandra Youkhana and
Luke Caspar Pearson**

Learning from Los Santos. A drawing exposing the different systems that regulate Los Santos and allow it to deviate from Los Angeles proper. You+Pea.

An allegorical, algorithmic architecture

Every day, millions of people travel to the virtual worlds of video games, diving into intense, escapist universes full of puzzles and adventure, grand battles and simulated societies. These synthetic worlds are built on software strikingly similar to those used in architecture, and for their players they are considered places where a collective experience of space creates social structures. Game worlds are tightly programmed, both computationally and architecturally. Their synthetic nature challenges normative conceptions of space and our interactions with it, expanding this to the whole environment, including its physics and rules. At the same time, games are generally entertainment media. We can consider their playful impulses to run counter to rhetoric in education, which tends to frame technology around the solving of grand problems, data collection or in the establishment of new design paradigms. The aesthetic experience of video games is immersive yet fragmented, inconsistent and often illogical. Yet this is precisely what attracts so many people to these virtual places.

As designers, we look at video games as both a mirror of our society's desires and a model for new ways of conceptualising, designing and even realising architecture that may have no prerogative to be built in the traditional sense. Games hint at unexplored possibilities for architectural design, including many concerns that have remained conceptual until now. It is this potential that we are working to uncover through our research and teaching. Around four years ago, we established a research studio called Videogame Urbanism at the Bartlett School of Architecture, that uses game engine technologies to conceptualise urban design projects. We saw games as a natural expansion of architecture's media landscape by embracing both engaging interactive systems and the widespread public familiarity with the medium. Our studio's research examines how the future of cities and the forces that regulate them can be communicated to new audiences and challenged through game technologies. Our pedagogic approach emphasises the use of speculative game scenarios drawn from real world research to incorporate systems, narrative and architectural theory into their construction. We develop projects that question the forces that shape contemporary urbanism, while also positioning the work in relation to the cultural and aesthetic influence of games. Following McKenzie Wark and Alexander Galloway's definition of video games as 'allegorithmic' structures,[1] where players have an allegorical relationship with an algorithm, making

Kintsugi City. Screenshot from a view of the game world showing all the compositional fragments within the game and paths of movements made by the player. Game by Yu Qi and Zhiyi (Zoey) Yang, Videogame Urbanism, 2018–19.

We see game systems as a barometer of public habits and politics in the information age.

games allows us to be self-reflexive about systems and simulations. In fact, Espen Aarseth argues that games are not merely unspecific allegories but 'allegories of space' that ultimately 'rely on their deviation from reality in order to make the illusion playable'.[2] Our research seeks to first unpick and then mobilise these deviations, because games are rarely, if ever, the 'true' simulations that other algorithmic design purports to be.

Mainstream architecture is already adopting game engine software, which is particularly prevalent in the field of VR visualisation. Both major game engine developers, Epic Games (Unreal Engine) and Unity Technologies, are producing architecture-specific versions of their platforms. However, we attempt to unpack their architectural qualities at a structural and cultural level that moves beyond the representation of a building for a client. Our students directly tackle the implicit relationship between encoded rules, interactive gestures and audio-visual representation that underpins the design of an interactive game environment. We argue that games utilise 'ironic computation',[3] where meaning is synthetically encoded and can be changed at a moment's notice. Games have been variously described as promoting 'failure',[4] 'repetition',[5] 'uncertainty'[6] and 'disunity',[7] none of which are terms that would find themselves as selling points for contemporary architecture or urbanism, but might more properly describe the friction of lived cities.

Towards a Videogame Urbanism

The intellectual basis of our studio's research reflects upon these properties of video games as a medium within architectural discourse. We discuss game structures as tools for learning and engagement, but also situate them in relation to the architectural canon of experimental projects from the past. It continues a history of architects fascinated by the impact of the culture industry, from the Smithsons collecting advertisements,[8] Archigram's comic books,[9] Venturi and Scott Brown's 'form analysis' of Las Vegas's pop architecture[10] or the morphological experiments of the 1990s using visual effects software. While popular discourse in universities is being driven towards high-level computer science – robotics, artificial intelligence and machine learning – we see game systems as a barometer of public habits and politics in the information age. At a point where algorithms are increasingly deified, Ian Bogost argues that by foregrounding their status as digital 'caricatures', games remind us that digital systems are messier and more

human than some would care to admit.[11] Alexander Galloway agrees, stating that games 'do not attempt to hide informatic control, they flaunt it'.[12] While the politics of the games industry is often called into question, the medium itself is well positioned for students to take a critical stance on all the systems being brought into architecture and urbanism.

Since its founding, Videogame Urbanism has produced over 50 playable games (and counting) of varying complexity and scope, from simple geometrical 'sandboxes' to complex urban-level simulations and networked multiplayer environments. Students generally arrive at the studio with no background in game design, game engine software or associated coding languages. As such, our pedagogic approach is centred around learning skills and critical theory through the production of design projects, the output of which are playable interactive gaming applications containing virtual environments. There is no one genre, viewpoint, aesthetic or system that defines video games, and likewise we attempt to avoid a single and fixed approach in our studio. For us, the most emblematic and overtly urban games, such as *SimCity* or *Cities: Skylines*, promote a top-down, Americanised version of urbanism[13] that should be challenged by games that allow people to experience cities through different eyes, with varying levels of agency and with different motives.

New eyes and agencies

Games are interactive and involve decision-making on the part of the player, which is why it is important to go beyond the view of the omniscient administrator of SimCity and investigate the agency of citizens. The Videogame Urbanism students who created *Carbon Neutral Living* (2018) used the structure of an RPG ('role playing game'), where players typically build a character whose expertise grows with repeated achievements. Rather than defining players through their mastery in swords and spells, the game was structured around maintaining their real-world carbon footprint. In the game, decisions made by the player in their 'everyday' life results in the opening or closing of parts of the game city as a provocation for how to incentivise an individual to change their lifestyle. How ecologically their character lived determined their access to urban amenities. Here, a game is instrumentalised to discuss the implications of urban living at a personal scale that causes the city to respond in real time.

We also look to challenge the 'god's eye' of designers as a visual position. In *Kintsugi* City (2019),

Carbon Neutral. Screenshot of a player replanting trees in the Olympic Park to offset their carbon footprint. Game by Yun Tie, Li Zhu and Zhaowei Zhu, Videogame Urbanism, 2017–18.

students created an alternative reading of Tokyo through the spatial protocols of Japanese art. By assembling fragmented objects through digital 'kintsugi' (the art of repairing broken pottery with gold leaf), a series of views through the city could be composed. Each of the hundreds of fragments in the game world was in constant communication with the others, to be manipulated by the player and reconciled through the flatness of the virtual camera. The game suggests that Eastern principles of art might be applied to our contemporary mapping and visualising software, which remains overwhelmingly Western in its outlook.

Games offer us the opportunity to make tools that provide direct feedback for designers and which non-designers can use, providing accessibility for people without the means or voice to participate in the typical processes of architectural design. The *Playable Planning Notice* (2017) responds to the UK planning system and its infamous laminated A4 notices placed on lamp posts. The description of potential changes is often either very technical or frustratingly vague, such as 'various works to various trees' or 'erection of a plinth and statue'. The game allows players to prototype their own subjective interpretations of real London planning notices through a building toolkit. While it has a visual resemblance to construction or city-building games, it is positioned as a social tool to increase the visibility of changes taking place within the city, far from the typical elevated overview of *SimCity*. These interactions could also take place on the site, as the paper notices would be replaced with smart technology that allowed citizens to view and interact with the proposals in real time.

All game worlds rely on time, in both seconds and computational 'frames', which means we can also challenge the temporality of decision-making and project both forwards and backwards. *Plug-In Tokyo* (2019) takes a series of dense urban sites in Tokyo and allows players to build structures through history as a vertical palimpsest, revealing the forces that shaped the city into its modern form. The history of architectural design is also key to this critique, and within the studio we have also reimagined precedent projects through games. A 2017 project, *Playing the Metropolis of Tomorrow*, challenged students to realise a set of unbuilt urban projects as game spaces. From Constant's *New Babylon*, to Archizoom's *No-Stop City*, Arata Isozaki's *City in the Air* and Walt Disney's EPCOT, students used game systems to critique each conceptual project. This introduced them to the agency of making games by revisiting and building upon a set of radical historical ideas of what urbanism could become through a new medium. Games emerged that explored living with Superstudio's *Continuous Monument* or revealed infrastructural failures behind Yakov Chernikhov's heroic *Architectural Fantasies*. An analysis of EPCOT as a self-contained urban 'bubble' led to a game where players could explore a spherical urbanism, while Kazimir Malevich's *Arkhitektons* informed a game shifting in scale from an art gallery to an infinitely expanding city. This approach also demonstrates that while our media may be new, the design of virtual worlds fits into a longer lineage of what we can call speculative architecture, where the boundaries of the discipline become expanded.

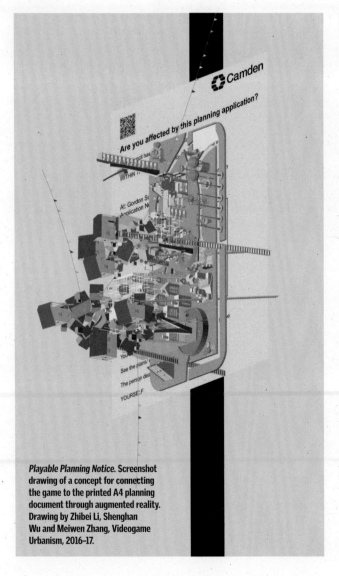

Playable Planning Notice. Screenshot drawing of a concept for connecting the game to the printed A4 planning document through augmented reality. Drawing by Zhibei Li, Shenghan Wu and Meiwen Zhang, Videogame Urbanism, 2016–17.

While our media may be new, the design of virtual worlds fits into a longer lineage of what we can call speculative architecture, where the boundaries of the discipline become expanded.

Beyond the Bubble.
A game exploring a spherical urbanism that has its own gravitational pull. Game by Yingying Zhu, Videogame Urbanism, 2017–18.

Semiotic Switch-Up.
Screenshot drawing of a constellation of objects within a casino which can be encoded with different behaviours by the player. Drawing by Yuting Pu and Zichun Yang, Videogame Urbanism, 2019–20.

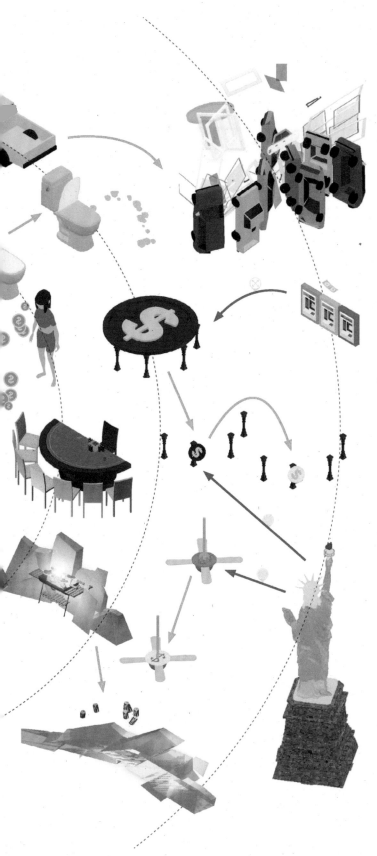

Situating the virtual

Site analysis in the more traditional sense is also a key component to the development of projects and allows us to situate the games in relation to real cities and their morphology. A 2019 project, *Playing the Machine Zone*, drew from Natasha Dow Schüll's writings on electronic gaming in Las Vegas as a counterpoint to Venturi and Scott Brown's assessment of the city as a language of 'communication over space',[14] which also arguably describes how video game environments work. Las Vegas provided the context for conversations about both the role of gaming (and gambling) at an urban scale and the role of Vegas's semiotics in relation to video game spaces. In the game *Semiotic Switch-Up* (2019), students designed a Vegas-style 'ergonomic labyrinth'[15] where every object, from a person to a sign or a craps dice, could have its properties transformed by the player. Navigating through the casino space, the semiotic connections between game objects and their affordances are changed in real time. This is an ideal example of ironic computation where the synthetic nature of game objects allows us to question design assumptions through severing and reconnecting meanings. In *Las Vegas Navigation* (2019), students built a set of AI agents that follow their own desires by wandering through a casino that would be updated and reconstructed around them by the player. In playing the game, the processes by which Las Vegas's seemingly uncontrolled spaces are structured would be brought to the fore.

As cities shut down during the COVID-19 pandemic in 2020, game worlds became elevated into social media proper, a way of people meeting and creating places and societies with minimal physical infrastructure. During this time, research has pivoted towards using networked online game environments as a collective digital urbanism. Networking adds another layer of complexity to the design of synthetic virtual worlds. The meaning, function and performance of objects must be communicated and synchronised between players. It poses interesting questions for our students, both technical and philosophical in nature. In the online game *OddWorld* (2020), players can alter the scale, appearance and behaviour of any object in the world, creating a free environment for building and teamwork, divided up into different zones. Players can collaborate to construct but also tie their designs to 'objectives' relating to urban forms made from all manner of objects within the game world. Here the meaning of objects becomes doubly layered because a player upending and scaling a 'bus' to become a 'building' must

OddWorld. Screenshot of an online game where players can manipulate the behaviour of objects in the world to collaboratively build structures together. Game by Yuting Pu and Zichun Yang, Videogame Urbanism, 2019–20.

also communicate that action to other users across the network. If this does not happen, the different worlds will diverge from one another. If games allow us to play with the meaning and affordances of entirely synthetic objects, networking adds another layer of complexity where the vagaries of connectivity can also alter an object's semiotic existence.

New practices of play

Videogame Urbanism studio is an extension of the research themes and methodologies we use in our design studio You+Pea. In order to understand the nature of contemporary game spaces, and the cultural and spatial 'impulses' they produce in players, our research involves the critical survey of commercial game worlds. In *Learning from Los Santos*, we built upon the 'form analysis' methods of Venturi and Scott Brown to analyse

Playing the Picturesque. View of the installation where visitors sit on a re-creation of one of the viewing benches designed into Nash's cottages for Blaise Hamlet, looking through the window into a game world beyond. You+Pea.

the digital counterpart to Los Angeles that forms the backdrop to the bestselling game *Grand Theft Auto V*.[16] Through cartographic mappings, photographic recordings and data mining the game, we can understand the ways Los Santos compresses and distorts LA around the logic of the game. It is in these moments Aarseth calls 'deviations' where the architecture (in our disciplinary sense) of game worlds can be found.

This research also leads us to draw parallels to other moments in architectural design that employ similar spatial strategies. *Playing the Picturesque*, a project exhibited at the RIBA Architecture Gallery in 2019, explores the symbiosis between picturesque architectural principles and the design of game worlds. By integrating video games into physical structures by means of projections and pressure pads, we extended the gallery into the realm of a virtual picturesque to demonstrate how contemporary game worlds often use similar design principles. The virtual interactions were triggered by the player's position and movement in the gallery, making them more accessible and mirroring how one would interact with a real picturesque landscape. Complementary strategies include routes that 'never let the foot travel the path of the eye',[17] an emphasis on the irregular and asymmetrical experience of space and 'sham' structures that appear complete from one angle but are revealed as flat facades or fragments on closer inspection. Establishing a one-to-one relationship between the viewer's body and the virtual space, we challenged the typical 'power-fantasy' narrative of amplified physical mobility and dexterity that is common to many games, instead embodying a gentle promenade to realise the virtual nature of the picturesque within the confines of a gallery. Picturesque designs were typically commissioned by the powerful, but we believe the construction of open virtual experiences can provide new collective experiences and reappraisals of historical hierarchies.

The *London Developers Toolkit 2.0* (2020) similarly deals with gaming in relation to urban power structures. In the game the player is responsible for building and then advertising a high-rise luxury development. The game has the loose appearance of a 'city-builder' but the mechanisms speak more specifically about the architectural imagery that accompanies flows of capital in the city. The player notionally operates as an architect for a pair of property developers, undertaking several tasks involving the gesture recognition of 'napkin sketches' and performing parametric calculations. Following this, a 'poor *Photoshop*' programme becomes accessible from

Consider a future where the disciplinary boundaries are so fluid that our distinction between projects realised physically or virtually disappears.

Playing the Picturesque.
View of the installation showing follies derived from Nash's architecture connecting to a virtual world designed around Windsor's Long Walk. You+Pea.

which players can design their own advertising images for their tower. The game is connected directly to a printer and a Twitter bot, which aggregates and promotes imagery from the game, connecting the virtual world to social media as a further contemporary equivalent to the Smithsons' impulse-creating ads. Designed through 'Minecraft-style' voxel modelling, the game has a blocky and childlike look. This cuteness has an agency, used to challenge 'from a position of playful vulnerability' and 'light-heartedly probe[s] the established ways in which we invoke power'.[18] Approaching architectural components in this way reinforces their 'gameness' and reiterates Bogost's claims about games and simulations as caricatures of reality because, as Ernst Gombrich argues, 'caricature, showing more of the essential, is truer than reality itself'.[19]

Games can be seen as tools, while at the same time being aware that their nature as entertainment without utility might indeed be their power as a medium. Students should be encouraged to produce sophisticated computational projects, yet they should also be asked to question why and how computers are used in architecture and urbanism today. If the architectural environments might seem cartoony, immature or kitsch, the sheer number of people embracing them as real and social places demands our attention. Graeme Kirkpatrick argues that 'all video games are a kind of opening up of the machine' that lever it out 'from the dominant historical narrative of "technological progress"'.[20] For architecture this means they provide a valuable reflection point as software and algorithms that make us question the nature of software and algorithms. It appears that this is where the key potential of using games as an architectural medium lies. We are able to synthesise virtual worlds, while also designing the messages and methods by which people can engage with them.

Videogame Urbanism asks architecture students to consider a future where the disciplinary boundaries are so fluid that our distinction between projects realised physically or virtually disappears. The future is already here, and we play it every day. It is tied to the re-emergence of ideas such as the decades old metaverse, the notion of an entire, secondary digital world. Contemporary discussions propose this could be made by connecting many smaller virtual worlds together through software protocols. For us, this is the right time to delve deeply into game environments as networked virtual urbanism and to build new forms of pedagogic practice ready for these new sites and contexts, to go beyond utopia and into new worlds proper.

London Developers
Toolkit 2.0. A drawing of
various game structures,
using a 'cute' voxel-style
as a way of subverting
development aesthetics.
You+Pea.

The future is already here, and we play it every day.

1 Wark, M., *Gamer Theory*, Harvard University Press, Cambridge MA and London, 2007, Note [030].
2 Aarseth, E., 'Allegories of Space. The Question of Spatiality in Computer Games' in *Cybertext Yearbook 2000*, M. Eskelinen and R. Koskimaa, Jyväskylän yliopisto, Jyväskylä (eds), p. 169.
3 Pearson, L., 'Architectures of Ironic Computation: How videogames blend representation, code and interfaces to offer new protocols for architectural experimentation' in *Inflection: Journal of the Melbourne School of Design*, vol. 3, 2016.
4 Juul, J., *The Art of Failure: An Essay on the Pain of Playing Video Games*, MIT Press, Cambridge MA and London, 2013.
5 Grodal, T., 'Stories for Eye, Ears, and Muscles: Video games, media, and embodied experiences' in *The Video Game Theory Reader*, eds M.J. P. Wolf and B. Perron, Routledge, New York, 2003, p. 148.
6 Costikyan, G., *Uncertainty in Games*, MIT Press, Cambridge MA and London, 2015.
7 Kirkpatrick, G., *Aesthetic Theory and the Videogame*, Manchester University Press, Manchester, 2011, p. 112.
8 Smithson, A. and Smithson, P., 'But Today We Collect Ads', *Ark Magazine*, no. 18, November 1956.
9 Archigram, *Archigram 4: Amazing Archigram / Zoom, Archigram*, London, 1964.
10 Izenour, S., Scott Brown, D. and Venturi, R., *Learning From Las Vegas*, MIT Press, Cambridge MA and London, 1977, p. xi.
11 Bogost, I., 'The Cathedral of Computation', *The Atlantic*, https://www.theatlantic.com/technology/archive/2015/01/the-cathedral-of-computation/384300/, 2015, accessed 30 July 2020.
12 Galloway, A.R., 'Playing the code: Allegories of control in Civilization', *Radical Philosophy*, 2004, https://www.radicalphilosophy.com/article/playing-the-code, accessed 20 July 2020.
13 Lobo, D.G., 'Playing with Urban Life: How SimCity Influences Planning Culture' in *Space Time Play: Computer Games, Architecture and Urbanism: The Next Level*, M. Böttger, F. von Borries, S.P. Walz (eds), Birkhäuser Verlag, Basel, 2007.
14 Izenour, S., Scott Brown, D. and Venturi, R., *Learning From Las Vegas*, MIT Press, Cambridge MA and London, 1972, p. 8.
15 Dow Schüll, N., *Addiction by Design: Machine Gambling in Las Vegas*, Princeton University Press, Princeton NJ, 2014, p. 54.
16 Rockstar North, *Grand Theft Auto V*, Rockstar Games, PS4/PC, 2015.
17 Macarthur, J., *The Picturesque: Architecture, Disgust and Other Irregularities*, Routledge, London and New York, 2007.
18 May, S., *The Power of Cute*, Princeton University Press, Princeton NJ, 2019.
19 Gombrich, E.H. (with Ernst Kris), 'The Principles of Caricature' in *British Journal of Medical Psychology*, Vol. 17, 1938, p. 319.
20 Kirkpatrick, G., *Aesthetic Theory and the Video Game*, Manchester University Press, Manchester, 2011, p. 9.

Evolving Design:

From Computer Tools to Generative Design Partners

Danil Nagy

Technological development has always impacted design and the way in which designers work. All design tools can be understood as technologies which continuously evolve to give designers new abilities to imagine and execute their design visions. The pencil a designer uses to take the ideas in their head and test them out on paper is a form of technology, as is the hammer that drives the nail into the wood that allows that idea to take shape.

By far the most impactful technological development of the last century was the arrival of the personal computer in the 1980s. Although computers had existed for several decades, their large size and expense kept them isolated in research labs and government institutions. The personal computer allowed everyday design practices to bring this technology into their studios and ushered in a new era of digital design tools that have fundamentally altered the way designers do their work.

While the earliest design software focused on replicating traditional drawing methods in a digital format, new tools have emerged that allow designers to use computers in totally new ways. Recently, there has been a growing interest in Artificial Intelligence (AI) technologies that allow the computer to make high-level decisions which automate aspects of the design process. Although these technologies are relatively new, there has been early success in an approach called Generative Design, which uses parametric design and optimisation algorithms to automatically derive high-performing design solutions based on a set of high-level goals and constraints.

Using these new digital tools requires learning new skills, such as geometry, probability theory and computer programming, which may not be familiar to many designers. But they also offer us new opportunities to design in better, more efficient ways, thus continuing the co-evolution of technology and design practice.

Designing with computers

The earliest design software focused on replicating traditional design methods in a digital form, replacing physical artefacts like drawings and models with virtual representations of points, lines and surfaces crafted in a virtual 2D or 3D environment. The digital format offered significant advantages over its physical counterparts, particularly when it came to modifying the design. If something needed to be changed, you could simply recall the relevant data from the computer's memory, modify it as necessary and save it back to memory. As a new digital medium, it created a revolution in the way designers work, allowing them to make faster changes to their design representations as their ideas evolved. However, it did not change representations from being static and needing to be manually crafted by the designer.

The next evolution of digital design tools came with Building Information Modelling (BIM), which allowed designers to construct their buildings virtually in a 3D environment using digital representations of

Comparison of a 2D drawing (left) where building elements are represented with lines and a BIM model (right) where building elements are represented with 3D objects.

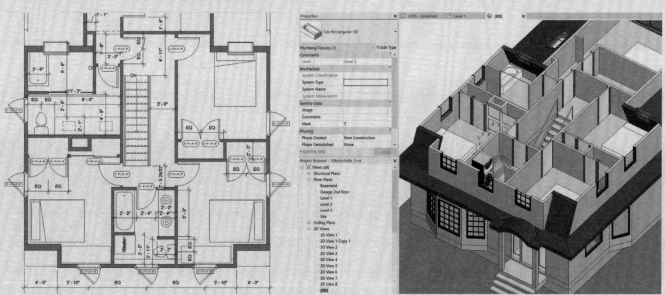

actual building components such as walls, floors, roofs and fixtures. BIM allowed designers to build their designs directly without worrying about coordinating a series of 2D drawings. It also saved time and reduced errors when making changes to the design, since the change only needed to be made once. All the required representations, such as 2D drawings, schedules and 3D renderings could be extracted automatically from the model.

Designing with algorithms

With traditional 3D modelling tools, including BIM, users create their designs by manually building them using a set of interactive tools. It is an intuitive way to work for most designers, but it often involves executing many repetitive tasks. Manual steps have to be repeated if changes to the model are required, which creates extra work.

The next major development in digital design tools arrived with the introduction of software that allowed users to create designs by encoding sets of instructions, called algorithms, directly into the computer. Algorithms tell the computer how to create the forms of a design, thus automating the manual process a designer would need to follow to create the design manually. To control how the algorithms are run, the user can expose a set of parameters, which are numerical values that control the way the instructions are carried out at a high level.

This software led to a new design practice called parametric design, shifting the focus from single, static designs to computational models based on algorithms that describe entire design concepts in a systematic way. Although developing such models can be more difficult and more time-consuming than simply drawing or modelling a static form, they offer significant advantages for designers. Because they represent the entire process for creating a particular design solution, they can automate many repetitive tasks. They also make implementing changes easier, since any part of the model can be adjusted and then the final result automatically regenerated.

Many digital design tools offer some support for parametric design, whether through scripting interfaces where users write algorithms using computer code or visual programming interfaces where users build algorithms using visual components in a drag-and-drop interface. Revit,[1] a widely used BIM software from Autodesk, supports both code-based programming in the C# language and visual programming through the plug-in Dynamo.[2] Rhino,[3] a popular 3D modelling software from McNeel, supports in-app scripting using a variety of languages, as well as visual programming through its plug-in Grasshopper.[4]

From parametric design to design automation

Although parametric design is still a relatively new concept, we are now witnessing the emergence of a new era of digital design tools that promise to automate our design process even further. This innovation is based on computer software that not only stores data and follows instructions but also makes high-level decisions that allows it to automate more complex cognitive tasks.

The technology behind this software has been researched for decades in the fields of Artificial Intelligence (AI), Machine Learning (ML) and Optimisation. Although software based on this technology is already in common use in many industries such as banking, insurance, manufacturing and logistics, it has been slow to develop in the design industry, mostly due to the personal and usually non-repetitive nature of design work.

These challenges have made it difficult for traditional software developers to create off-the-shelf automation software that can appeal to the needs and approaches of every designer.

Recently, however, designers have started to use advanced techniques like scripting and computer programming to work with these technologies directly and build their own tools and workflows that suit their needs and interests. One promising approach is Generative Design, in which advanced optimisation algorithms are used to automatically derive optimal design solutions from parametric models based on a set of high-level goals and constraints.

The process starts by creating a parametric model which defines a range of solutions to a certain design problem and computes a set of metrics that can be used to judge the performance of each solution. The model is then connected to an optimisation algorithm, which explores the model's parameter values and figures out which settings produce the best overall designs based on the specified metrics. Several tools now exist to support the Generative Design process, mostly in the form of add-ins or plug-ins, which add the optimisation component to an existing parametric design or modelling software. Some examples include:
- Galapagos, a simple optimisation tool included with Grasshopper
- Refinery,[5] which offers optimisation for Dynamo, and
- Discover,[6] a Grasshopper plug-in created by the author.

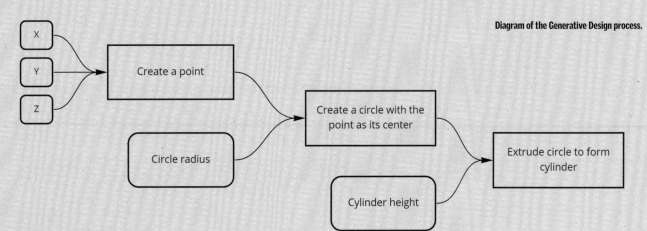

A cylinder defined using a set of parameters and geometric operations in Grasshopper (left) and visualised in Rhino (right).

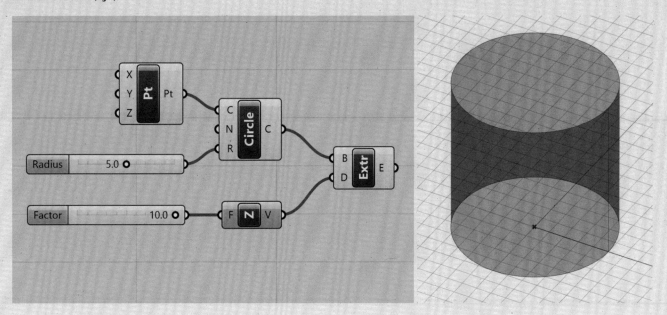

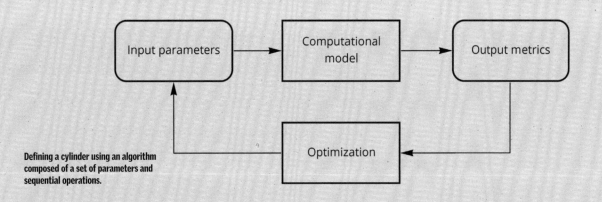

Defining a cylinder using an algorithm composed of a set of parameters and sequential operations.

Designing good design spaces

A common misconception about Generative Design is that it automates the design process when, in fact, it only automates the process of discovering good solutions from a given parametric model. Since the optimisation algorithm can only produce designs which can be created by the model, the quality of the outcome depends heavily on the quality of the model, the range of solutions it can produce and the information we give to the algorithm to steer it in the right direction. Creating such a model requires a significant amount of effort and a new set of skills for the designer, including an understanding of maths (particularly geometry) and computer programming. More importantly, it requires a new way of thinking about the design process and a new approach to how we collaborate with our digital tools.

When designing a parametric model for optimisation, you must imagine not just a single outcome but all the possible options produced by that model at once. To help visualise this range of possibilities, imagine the model defining a conceptual design space of possible solutions, with each parameter defining a dimension of that space and each possible solution occurring somewhere in that space. The goal of the optimisation algorithm is to search through this design space and discover the best-performing designs contained within. The goal of the designer, on the other hand, is to design good design spaces which define a rich set of possible options for the algorithm to explore.

Like any design task, designing good design spaces is a creative and challenging exercise which requires a great deal of technical knowledge, experience and intuition to do well. Compared to a traditional design process, this approach can also pose extra challenges for the designer. Instead of focusing on creating a single design object, the designer needs to think along a multitude of dimensions at once and envision a single system that creates not just one but a potentially infinite set of design solutions. While creating good design requires the designer to think of form, function, composition and beauty, creating good design spaces requires the designer to think in terms of geometries, rules and systems.

Working this way requires learning new technical skills, which may be intimidating for many designers. Knowing these methods, however, will bring great benefits in the reduction of manual repetitive work and can even lead the designer to discover novel solutions that they would not have come up with on their own.

> **The designer needs to think along a multitude of dimensions at once and envision a single system that creates not just one but a potentially infinite set of design solutions.**

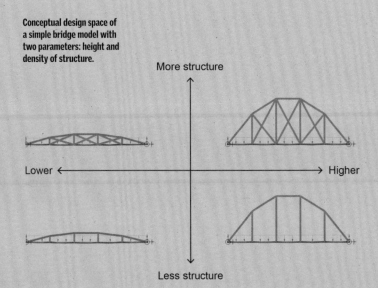

Conceptual design space of a simple bridge model with two parameters: height and density of structure.

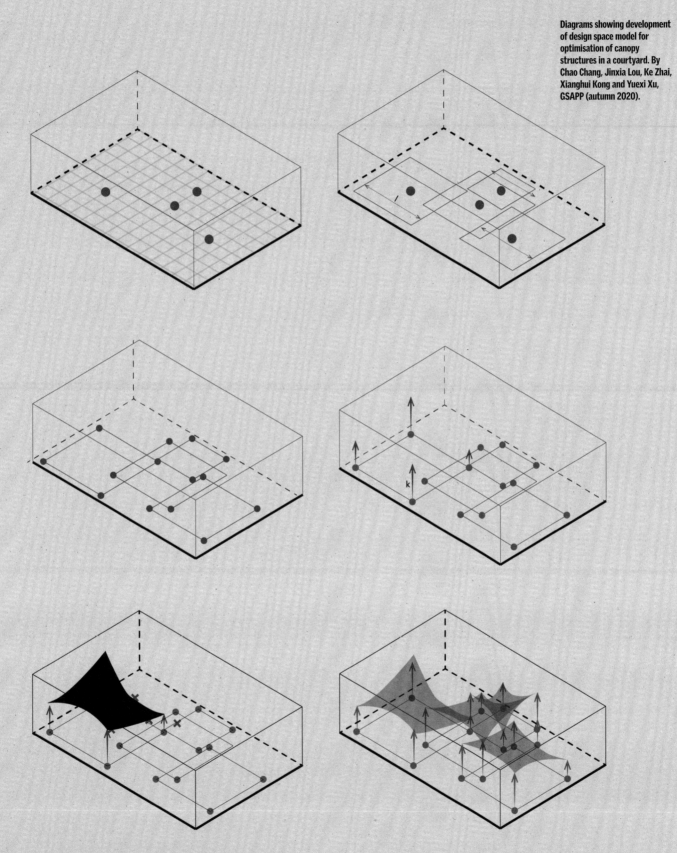

Diagrams showing development of design space model for optimisation of canopy structures in a courtyard. By Chao Chang, Jinxia Lou, Ke Zhai, Xianghui Kong and Yuexi Xu, GSAPP (autumn 2020).

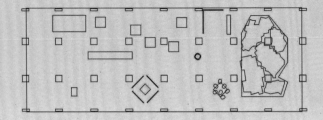

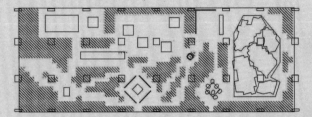

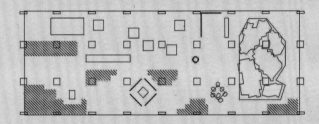

Process for analysing foot traffic within a gallery and optimising a roof structure based on active and inactive pedestrian zones. By Chen Yang, Liwei Guo, Nanjia Jiang, Tianheng Xu and Wan-Hsuan Kung, GSAPP (autumn 2020).

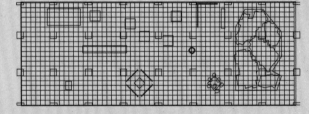

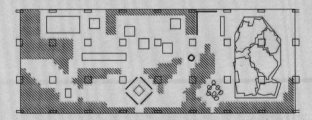

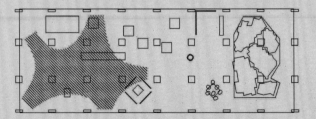

As an approach, it also transforms the software from a tool used by the designer to represent a design to a true partner in the design process. In this new process, the human designer is responsible for describing the approach to solving a problem, while the computer assistant iterates through many different options until it finds the best one.

Designing with parameters

One of the biggest challenges of designing good design spaces is exposing the right set of parameters in the parametric model. A model's parameters establish the dimensions of the design space and control the variety of options it can produce. Since the optimisation algorithm will not know anything about how the model works or the problem you are trying to solve, its ability to find good solutions depends heavily on the number and type of parameters that you use. Although parameters are always defined by numbers, there are three unique parameter types that give us different ways of controlling the model:

1. *Continuous parameter* defines a number within a certain range and should be used to control aspects of the model which vary smoothly. For example, the dimensions of a building would make good

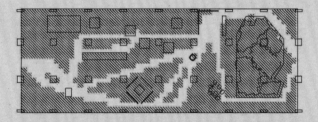

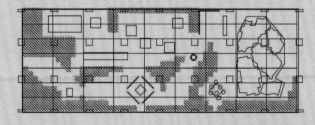

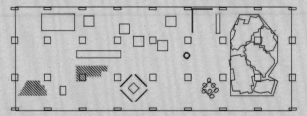

continuous parameters, since you can set a minimum and maximum value and explore different dimensions within that range.

2. *Categorical parameter* defines a selection from a discrete set of options and should be used in cases where the possible options do not have a continuous relationship to each other. For example, if you have a model of a building facade and want to control the type of window for each element, you can represent that choice with a categorical parameter.

3. *Sequence parameter* defines a set of values that describe a unique ordering of a set of elements. This parameter can be useful to control the order of a process or the assignment of certain elements to each other. For example, if you want to create a model defining rooms in a floor plan, you can use a sequence parameter to assign programs to those rooms in different combinations.

The type of parameter you choose depends on the nature of your model and how you want to control it. Understanding your options is important, however, because the number and type of parameters you choose will affect the variety of options that can be produced by the model and thus the range of options the algorithm can explore.

Designing with objectives and constraints

Input parameters give the optimisation algorithm a way to create a variety of design options from a parametric design space model. To make sure it generates the best design solutions, you also need to give the algorithm a way to evaluate each option and determine which options are better than others. To do this you need to describe the goals of the problem using output metrics that can be calculated for each option within the design space. Most optimisation algorithms support two kinds of output metrics:

1. *Objectives* describe design goals in relative terms, allowing you to measure the performance of each design relative to other designs. When you set an output metric as an objective, you must tell the algorithm whether you would like that value to be minimised (made as low as possible) or maximised (made as high as possible).

2. *Constraints* define the goals in absolute terms, as specific criteria that an output metric must meet for the design to be deemed valid. Unlike objectives, constraints do not describe relative performance between different design options. All options that meet a certain constraint are considered equally good, and any options that break a constraint are considered equally invalid.

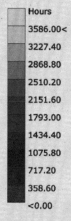

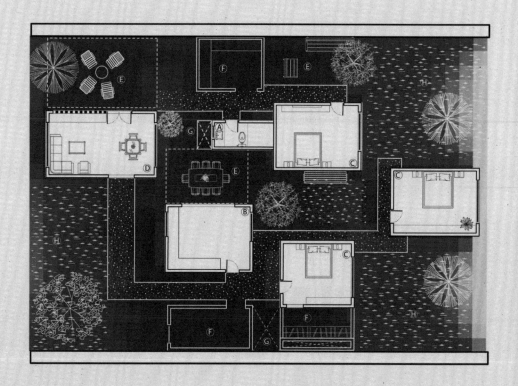

Optimising the layout of residential spaces based on program and daylight analysis. By Brennan Heyward, Cong Diao Jr, Joachym Joab, Kshama Daftary and Tianyuan Deng, GSAPP (autumn 2020).

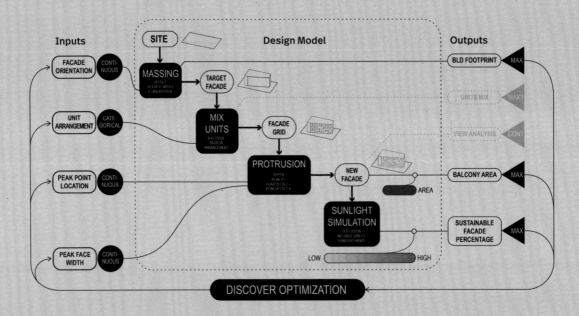

Objectives and constraints provide two different ways to communicate design goals to an optimisation algorithm. Given a set of objectives and constraints, the optimisation algorithm will try to create designs that meet the conditions of all constraints while pushing the objective values to be as optimal as possible.

Case Study

Generative Design course at Columbia University

Generative Design is an advanced technical elective course taught by the author at Columbia University's Graduate School of Architecture, Planning and Preservation (GSAPP).[7] The course introduces students to the concept of Generative Design to teach them practical skills in developing parametric models that represent good design spaces for optimisation. Using a variety of tools, students work collaboratively in groups to develop projects that test the Generative Design workflow on real-life design problems.

For each project, students must:
· choose a specific design problem
· design a parametric model that describes a design space of possible solutions to the problem, and
· optimise the model to find the best solutions.

Since the inception of the course, students have been encouraged to apply a Generative Design approach to many design problems at a variety of scales, including furniture, facades, buildings and even entire city blocks. Through this project-based learning approach, it has been observed over several years that students not only learn the technical skills and methods for creating good design space models, but they also, in parallel, naturally experiment and develop new concepts for how automation technologies can be incorporated into their creative design process. They therefore expand the possibilities afforded beyond the existing into potential future teaching research development and innovation, both within academia and real-world application.

Students learn a more systematic approach to design, centred on developing design concepts algorithmically and communicating them to a computer using both visual and text-based programming. The Generative Design approach seems to force students to increasingly think through their design concepts more rigorously and break complex problems down into smaller components that can be programmed into a computer.

The project-based experimentation and systemic approach developed in the evolution of this course demonstrates a new way for students to think about design and the future role of their digital tools, from tools that aid a manual design process to creative partners that actively participate in it.

Conclusion

Generative Design is just the latest in a long evolution of digital design technologies that have emerged since the arrival of the personal computer in the 1980s. Like any technology, these tools unlock new potentials in our design process, allowing us to design better and in better ways. Unlike previous developments in design software,

A new generation of computational designers who leverage AI and automation to do great design in a better, faster and more efficient way.

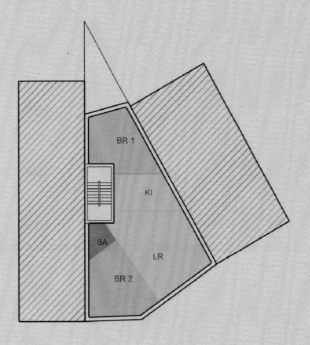

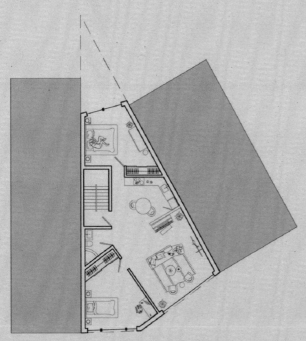

however, this latest shift promises to fundamentally change the role of the computer in the design process from a passive storer of information to an active participant. This has led many to speculate on whether computers will one day fully automate the design process, making human designers less important or even completely unnecessary.

Although modern software can now automate many of the routine aspects of the design process, human beings are still responsible for programming the rules and algorithms that the computer follows to accomplish its goals. Thus, there is a serious risk that these technologies will standardise the process and outcomes of design in a negative way, especially if they are placed outside the hands of designers. Generative Design suggests a different path, not of technology automating the design process but of a new generation of computational designers who leverage AI and automation to do great design in a better, faster and more efficient way. By combining the best of both human and computer intelligence, these designers will define the next era of design practice, evolving as ever in tandem with technological innovation.

1   https://www.autodesk.com/products/revit/overview
2   https://www.autodesk.com/products/dynamo-studio/overview
3   https://www.rhino3d.com/
4   https://www.grasshopper3d.com/
5   https://www.generativedesign.org/
6   https://colidescope.github.io/discover/
7   https://medium.com/generative-design-course

Entering a Bio-Based Material Paradigm:

**Probing Advanced
Computational Methods for
a Shift in Material Thinking**

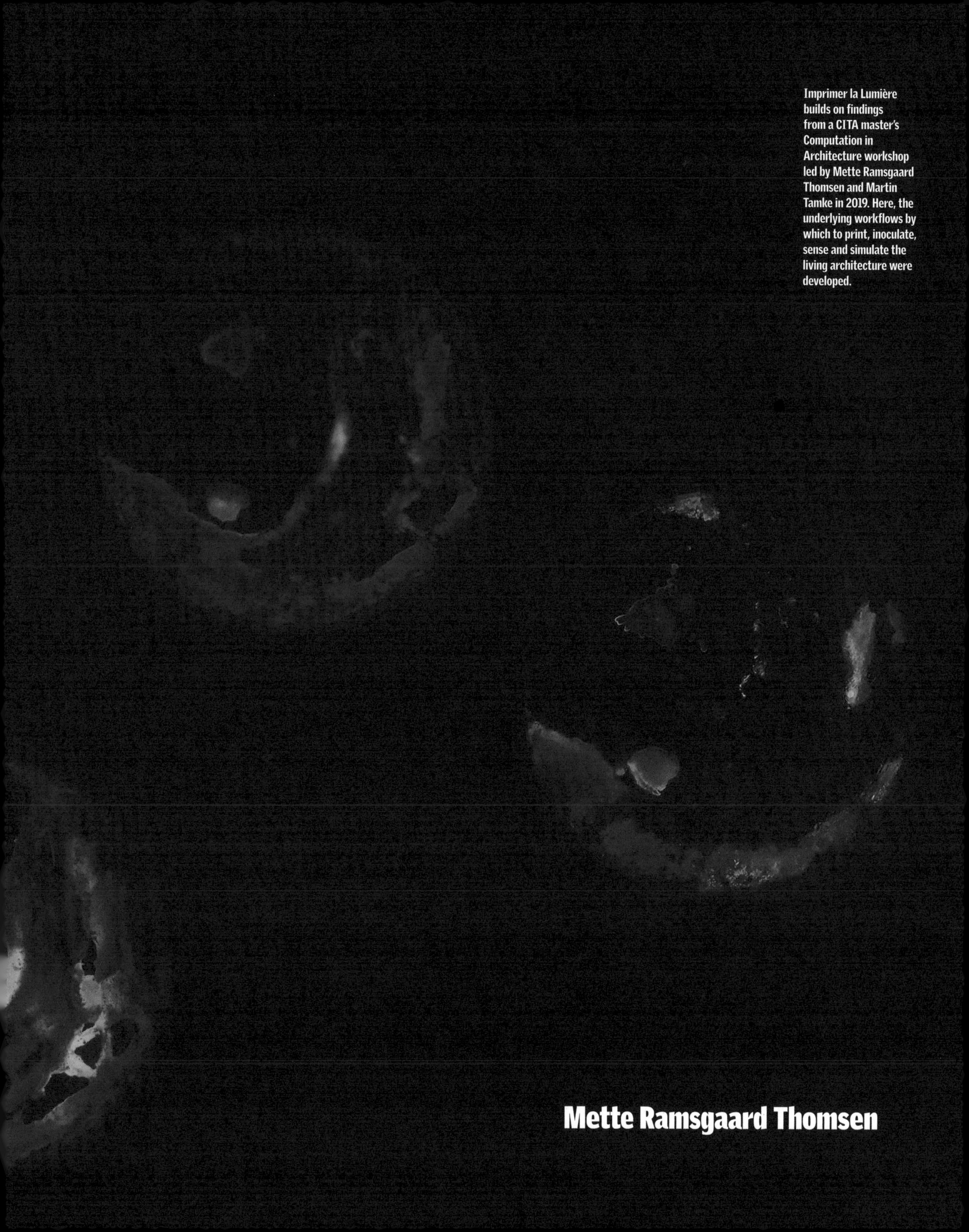

Imprimer la Lumière
builds on findings
from a CITA master's
Computation in
Architecture workshop
led by Mette Ramsgaard
Thomsen and Martin
Tamke in 2019. Here, the
underlying workflows by
which to print, inoculate,
sense and simulate the
living architecture were
developed.

Mette Ramsgaard Thomsen

How can we rethink sustainable building practices through a bio-based material paradigm? The following article posits that to truly enter a sustainable building culture we have to fundamentally question:
· what materials are, and
· how we work with them.

By challenging the perception of our built environment as inert, it argues that a new conceptualisation of bio-based materials will allow for sustainable, renewable, non-toxic, in-part biodegradable, versatile and reliable solutions to the growing shortage of building materials. It identifies that the key impediments to this transition lie with an inherent inability to represent and therefore fully conceptualise and, in turn, operationalise the complexity, behaviours and lifespans of bio-based materials. If current representational paradigms – and the digitised design and fabrication practices that they entail – presume homogeneity of structure, predictability of behaviour and stable longevity in material systems, thus advancing an essentially inert design space, the proposal here is of a material paradigm shaped by the dynamic behaviour of a bio-based material practice. The following presents a framework for bio-design in architecture of:
· the harvested
· the designed, and
· the living
that encompasses the breadth of bio-based material practice. The aim is to discuss how bio-design necessitates a rethinking of our representational practices that can capture, design and steer the dynamic transformations that biological materials undergo, their life cycles, evolution and decay.

A bio-based material paradigm
Construction is notoriously intense in its use of materials. In Europe alone, 30–50% of total material use goes into construction, with 65% of this being aggregates and 20% metals.[1] By predominantly depending on a small subset of materials extracted from the geosphere, we have created a global material crisis with severe socio-political, economic and environmental repercussions. Entering a bio-based material paradigm characterised by renewable resources and environmental circularity provides architecture and the built environment with new perspectives. However, set within this paradigm lie embedded differences in the ways in which materials perform within a structural system and over time fundamentally challenge how architecture is thought about, designed and produced.

The new narratives of bio-design and a circular bio-economy present alternative pathways to a sustainable future. By advancing the fabrication of products and energy made from the biosphere, they present a material paradigm of carbon-neutrality, renewability and circularity.[2] However, despite well-organised calls for action maturing into local legislation and an attentive profession, architecture is proving hesitant in this transition.[3]

Bio-based materials are fundamentally different to current building materials. If industrialised materials are designed to be homogeneous, static and stable, enabling certification and ensuring durability, bio-based materials are essentially transformational, evolving in time and across process. Shaped by growth cycles and formed by their environment, bio-based materials are characterised by their complex heterogeneity, unpredictable behaviours and the high degree of interdependence their life cycles embody. As instantiations of biomass, bio-based materials are part of a global steady-state system in which inflow is exactly balanced with outflow.[4] As such, their primary characteristic is their embedded temporality and inevitable decay. Present agendas share the ambition to produce materials that replicate the existing ones in performance and durability. However, this endeavour to facilitate continuity in our perception of what materials do, how they are employed and what their performances are, creates a false premise that limits our understanding of their design, fabrication and deployment.

Designing for and with bio-based materials challenges the fundamental value proposition of architecture. Instead of predicating design on an idealisation of permanence and the insurance of durability, we must expand our conception of material lifespan and find new ways of engaging practices of maintenance and intervention. Our aim is to propose a holistic conceptualisation of bio-based material practice that allows for carbon-neutral, renewable and materially optimised solutions to the growing shortage of building materials. By radically contesting the architectural axiom of firmitas, we propose a new representational framework to capture, design and steer the dynamic transformations that biological materials undergo – their life cycles, evolution and decay. This challenges our preconception design agency as restricted to the traditional cut-off point of building completion, instead incorporating a speculation on how new practices of continual construction, in which buildings are continually built and rebuilt as their materials are depleted or decay,

RawLam: experimenting with performance-based strategic grading of material resource. The composition of the beam element strategically places high-quality lamella around high-stress areas (surfaces, joints and end points) and low-quality lamella in low stress areas (interior).

The new narratives of bio-design and a circular bio-economy present alternative pathways to a sustainable future.

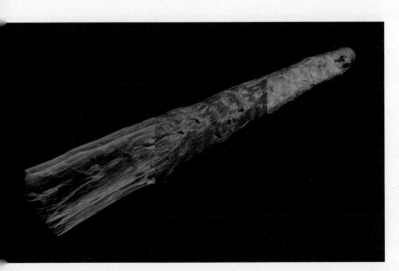

can recast the short lifespans of bio-based materials as effective properties of a new sustainable practice. It asks: how can we change the way we conceptualise bio-based materials – their performance, composition and durability – to allow a rethinking of sustainable design practice? Further, how can this new conceptualisation effect a transformation in construction practices?

Form-finding in a bio-based material paradigm

Where form-finding methods in computation design, consider material behaviour, they do not conceptualise and represent the lifespans of materials and are therefore equally linked to an idealisation of architecture as inert, and building completion as a cut-off point to design agency as their industrial counterparts. The field of bio-design is still only emerging. The conceptualisation of a bio-based materiality is being discussed through highly profiled networks[5] and curated exhibitions.[6] Maker-based designer-led interest is driving a breadth of new products[7] and the tie to traditional bio-based materials, such as timber, is cemented. However, there is no shared model. The field remains fragmented and characterised by disparate agendas engaging different communities of very diverse degrees of maturity, inconsistent goals and methodologies. In architecture, this breadth can be described across three material perspectives that engage different theoretical, technological and historical positions.

The harvested

The practices of building with timber, bamboo, reeds or straw are established in building history. The culture of seeding, growing and harvesting bio-based materials for architectural structures dates back millennia,[8] and timber elements themselves can last over a thousand years.[9] Harvested materials are understood as central to the future of sustainable building practice. The arrival of new harvested bio-based materials, such as mass timber, structural bamboo, straw bale fillers or hemp-based insulation panels, is changing how we think about architectural construction as part of an inherently renewable material practice offering new performances, such as embedded abilities to biodegrade and act like a carbon sink. Further investigations into mycelium bio materials, in which controlled processes of inoculating, feeding and arresting fungal organisms bio-based material stock to grow designed building blocks[10] or acoustic isolation materials,[11] are allowing us to rethink a future of manufacture that combines industrial processes of control with living technologies.

At the Centre for Information Technology and Architecture (CITA), 'RawLam' is an example of a harvested bio-based material. It examines how advanced computed tomographic imaging can be used to optimise the deployment of differentiated timber qualities in glulam (glued laminated) structures. The project examines the digital design chain in timber construction and develops new systems for design, integrating advanced imaging gained from the sawmill industry. The image stacks are analysed for density, fibre direction, wane and knots and interfaced with the design environment to directly map design performance requirements with available resource qualities. This allows us to strategically employ high- and low-quality timber within the glulam elements at a higher resolution than existing methods of timber grading.

In student workshops at the CITA master's Computation in Architecture, participants explored how varying the material composition and arrangement of glulam blanks can steer material performance of structural elements. The experiments were supported by the creation of 3D scan-based verification methods that map design intent to fabricated element and adjust connection detailing.

To capture and harness material variance is highly complex.[12] To truly understand and design with inherent material properties of harvested materials, architecture needs to:
· develop the ability to capture the heterogeneity of the single material
· represent its complexity
· predict its structural performance and associated behaviours, and
· interface these with novel models of design and fabrication for the optimisation of material deployment.

The designed

Bio-polymers, made from biomass such as starch, cellulose, gums[13] or microorganisms,[14] are currently attracting attention as ecologically sound alternatives to petroleum-based polymers.[15] Architectural research into bio-polymers focuses mainly on replacing glass fibre-reinforced plastics for panels or structural frames with bio-composites.[16] Slurry-based bio-polymers have been tested in different compositions.[17] However, the success of these investigations is still compromised by the difficulty of aligning material performance and durability with building criteria. Despite having preceded petroleum-based polymers in the 1940s and 1950s,[18] bio-

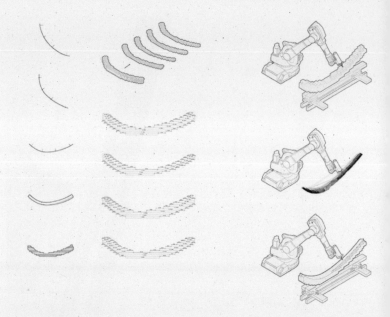

The CITA master's Computation in Architecture glulam workshop, led by Tom Svilans with Martin Tamke and Mette Ramsgaard Thomsen, explored the creation of novel glulam blank types by varying material parameters and robotically steered processing.

03

Student work exploring bio-polymer extrusion for architecture, understanding how material composition and print parameters affect each other. CITA master's Computation in Architecture 2019.

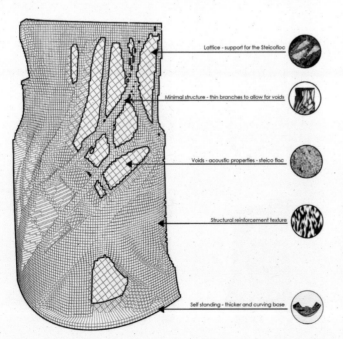

Lattice - support for the Steicofloc

Minimal structure - thin branches to allow for voids

Voids - acoustic properties - steico floc

Structural reinforcement texture

Self standing - thicker and curving base

polymers are not readily appropriated in architecture. Bio-polymers have low mechanical strength, shorter durability, less chemical inertness and demonstrate complex behaviours during fabrication.[19] A particular dimension of this is their high degree of volatility. Slurry-based bio-polymer composites are predicated on water-based recipes that make the composites formable but also entail large-scale anisotropic structural and dimensional transformations as the water content evaporates during curing. Designing with bio-polymers composites therefore necessitates the modelling of the transformational behaviours and *temporal plasticity*.

At CITA, 'Predictive Response' is an example of a designed bio-based material. It examines the design and steering of 3D-printed graded cellulose reinforced biopolymers for architecture. The project develops new methods for extruding bio-polymer slurries and examines how their material composition can be characterised and steered through processes of material variation and graded deposition. It investigates how sense data-tracking the temporal deformation of the material in time during curing can be interfaced with predictive models to forecast material performance.

The project was explored through student workshops in CITA master's Computation in Architecture course, in which 3D extruded elements were tested in respect to architectural performance criteria, including structural and acoustic performance.

To truly exploit the potential of bio-based polymers, we need new models that harness the limited duration and volatility of bio-polymers as active design parameters rather than failures. We need to interface material characterisation with fabrication to understand material performance and develop means of conceptualising how parameters affect and change material lifespan.

The living

The emergence of new bio-technology platforms is allowing novel explorations of an architecture embodying the living.[20] New perspectives are aspiring to position architecture as the host of functionalised biological organisms that metabolise and transform materials or produce energy. A particular trajectory within the field is the ambition to design with living systems as a direct part of the architectural body, functionalising them through their ability to metabolise and transform agents or produce energy. The aspiration of integrating living microbial fuel cells into the built environment – to use human waste for distributed energy production,[21] building facades hosting mosses or algae to purify air[22] or dedicated bio factories producing heat or nutrition[23] – define ways of understanding the building as the host of a new set of organisms that extend a building's performances.

Living materials are manifestly different to produce, manipulate and administer than inert materials. They are heterogeneous in that they are embedded in media, like hydrogel, that act as nourishing and oxygenating hosts. As the organism propagates, the media is depleted, changing material living conditions dynamically in time, necessitating either practices of nourishing or replacing material on an ongoing basis. Living materials are furthermore temporally plastic, expressing changing behaviours as they react to environmental changes such as circadian rhythms, temperature, humidity and the presence of other species. Finally, they embody 'different kinds of lifespans' as practices of cryobiology (freezing and thawing of living things) and cell synchrony (artificial manipulation of cycles of cell division) allow us to change the way organisms are grown, exchanged and stored.[24]

At CITA, Imprimer la Lumière is an example of a living bio-based material. It examines the use of light-emitting bacteria as architectural materiality.

Imprimer la Lumière, collaboration between CITA and Soft Matters Group, ENSAD. Bioluminescent bacteria are inoculated into 3D-printed agar medium structures, creating an illuminating micro-architecture. Simulations speculate on the propagation of bacteria and therefore light source through the material.

Imprimer la Lumière builds on findings from a CITA master's Computation in Architecture workshop led by Mette Ramsgaard Thomsen and Martin Tamke in 2019. Here, the underlying workflows by which to print, inoculate, sense and simulate the living architecture were developed.

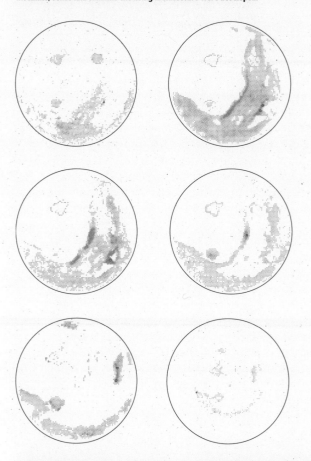

By investigating the 3D printing of bioluminescent bacteria, we question how architecture can be host for an ecology of species in sympoetic coexistence. Widely occurring in marine life, some mushrooms and insects, bioluminescent bacteria emit low light levels as part of their living metabolism. Imprimer la Lumière examines how these metabolisms can be incorporated into a living architecture and how their performances can be programmed as part of the architectural environment. Here, the distinction lies between the 3D-printed vessel and the biological mass, by producing the containment system as a gradable biomaterial. This can be locally adjusted to serve either as nutrition or as a shell for the bacteria, elongating longevity of the bacteria and providing a rich ground for design, where the distribution of bioluminescence across the object and space can be precisely specified.

In CITA master's Computation in Architecture, we have explored the metabolisms of bioluminescent bacteria developing methods of simulating their propagation through their host medium. These early studies examine how material life cycles can intersect with the temporality and programme of architectural structures.

Fungar, an EU FET collaboration in which CITA takes part and is led by Phil Ayres, models the topology of living fungal networks in order to control and functionalise them.

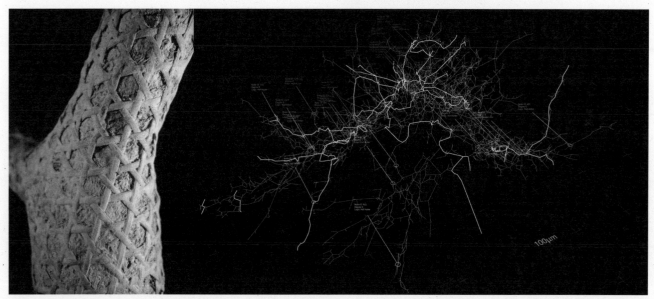

A second project at CITA, examining the idea of a living architecture, is the Fungal Architectures research project, which seeks to develop a fully integrated structural and computational living substrate using fungal mycelium for the purpose of growing architecture. The project is targeting two principle technological breakthroughs:
1. Growing monolithic structures at metre-length scales.
2. Functionalising the mycelium network to act as a computing/sensing device.

A construction approach is being developed that employs Kagome (tri-axial) weaving to create a combined reinforcement and stay-in-place formwork for growing mycelium composites. At a different scale, mycelium networks are being analysed to determine how their topology might be functionalised towards computational tasks.

Designing with living organisms necessitates the modelling of their performances as they react to environmental stimuli, their life cycles, propagation and decay, as well as the temporal transformation of their media as nutrition becomes exhausted. To understand, curate and manage living materials, we need to devise ways of predicting their performance as well as their propagation and lifespan as part of a creative architectural design process. Computational design and the integration of sensing of material systems as dynamic constructs offers means by which such methods can be instrumentalised and interfaced with design.

Conclusion: designing for the living

In order to enter a bio-based material paradigm, we need to develop new architectural representations that capture, design and steer the properties of bio-based materials, their heterogeneity, temporal plasticity and living performances, interface their fabrication and track a situated understanding of their behaviour over their lifespans. To imagine a future of a bio-based architecture we need new models of representation that capture environmental flux and the way the presence of inhabitants and climate afford changes in humidity, temperature and exposure, thereby affecting the composition, behaviour and growth cycles of living materials. This fundamentally changes the way we understand the interaction between inhabitant and edifice, necessitating a reorientation of architecture and disrupting the foundations of standardisation and durability. Instead, we need to develop ways to conceptualise and represent the lifespans of building materials and the dynamic processes that lead to their decay, reinforcement or evolution. We need to understand design action as a continually calibrated response to environmental change, embedding conceptions of circularity and feedback between processes of design and the evolving performance of transformative materials that breaches the cut-off point of architectural design agency as aligned building completion.

Acknowledgements
'RawLam' is developed as part of the Material Imagination project in collaboration with Aarhus School of Architecture. 'Predictive Response' is a collaboration between the CITA, the Technical University of Denmark and the University of Reading. Both are funded by the Independent Research Fund Denmark. Imprimer la Lumière is a collaboration between CITA and the Soft Matters Group at ENSAD: École nationale supérieure des Arts Décoratifs in Paris. Fungal Architectures is a collaboration between CITA, Computer Science/Biophysics (University of West England), Mycology (University of Utrecht) and industry experts in mycelium-based technologies (MOGU). The project is funded by the European Union's Horizon 2020 research and innovation programme (grant agreement No. 858132). Participating students: Harry Clover, Darius Narmontas, Fabian Puller, Enrico Pontello, Erik Lima, Mads Brandt, Emil Buchwald, Nicholas Mostovac, Sebastian Gatz, Teodor Petrov, Lina Baciuskaite, Jonas Mortensen, Gina Perier, Claudia Schmidt, Brian Cheung, Johan Pedersen, David Schwarzman, Mariel Dougoud, Luca Breseghello, Gabriella Rossi, Pamela Clover, Cepide Garivani, Daoyuan Zhu, Jamie Walker, Leonardo Castaman, Stian Holte, Agnes Hekla, Søren Tobias Jørgensen, Bartlomiej Markowski, Ao Tan, Ruxandra Chiujdea, Manish Bilore, Andrea Recinska, Aadil Amla, Yu Chen, Dominykas Savickas, Aroni Roy,Tessira Reyes Crawford, Claudia Colmo, Izabella Banas, Kawtar Al Akel,Ke Lin, Nikhila Vedula, Youcheng Li, Carolin Feldmann.

To understand, curate and manage living materials, we need to devise ways of predicting their performance as well as their propagation and lifespan as part of a creative architectural design process.

1 Herczeg, M., McKinnon, D., Milios, L., Bakas, I., Klaassens, E., Svatikova, K., et. al., 'Resource efficiency in the building sector', Final Report, 2014.

2 Hebel, D.E. and Heisel, F., 'Cultivated Building Materials', in *Cultivated Building Materials: industrialised natural resources for architecture and construction*, Birkhäuser, Basel, 2017, pp. 8–15. Blok, R., Kuit, B., Schröder, T. and Teuffel, P., 'Bio-based construction materials for a sustainable future', 2019 IABSE Congress New York City: The Evolving Metropolis, International Association for Bridge and Structural Engineering (IABSE), 2019, p. 1–7.

3 See 'A sustainable bioeconomy for Europe: Strengthening the connection between economy, society and the environment', Brussels: European Commission, 2018. Fund, C., El-Chichakli, B. and Patermann, C., 'Bioeconomy Policy (Part III)', *Update Report of National Strategies around the World*, German Bioeconomy Council, 2018. Sandak, A., Sandak, J., Brzezicki, M. and Kutnar, A., 'Biomaterials for Building Skins', in A. Sandak, J. Sandak, M. Brzezicki and A. Kutnar (eds), *Bio-based Building Skin*, Springer, Singapore, 2019, pp. 27–64. King, B., *The New Carbon Architecture: Building to Cool the Climate*, New Society Publishers, Gabriola, 2017.

4 Needham, A.E., *The Uniqueness of Biological Materials*, Pergamon Press, Oxford, 1965.

5 Network: Living Arcitecture Systems Group 2016. http://livingarchitecturesystems.com/

6 Myers, W. and Antonelli, P., *Bio Design: Nature*, Thames and Hudson, London, 2012. Brayer, M-A., Zeitoun, O., Papapetros, S., Bianchini, S., Quinz, E. and Collet, C., 'MUTATIONS/CRÉATIONS: LA FABRIQUE DU VIVANT', Centre Pompidou, Editions HYX, 2019. C. Collet, 'Alive: New Design Frontiers', EDF Foundation, 2013.

7 Bernabei, R. and Power, J., 'Living Designs: Biomimetic and Biohybrid Systems', 5th International Conference, *Living Machines*, Springer International Publishing, Switzerland, 2016, pp. 40–47. Karana, E., Blauwhoff, D., Hultink, E-J. and Camere, S., 'When the material grows: A case study on designing (with) mycelium-based materials', *International Journal of Design*, 12, 2018. Majumdar, P., Karana, E,. Sonneveld, M.H., Giaccardi, E., Nimkulrat, N., Niedderer, K., et. al., 'The Plastic Bakery: A Case of Material Driven Design', *Proceedings of the International Conference of the DRS Special Interest Group on Experiential Knowledge and Emerging Materials*, 2017, pp. 116–28.

8 Brunskill, A., 'Mesolithic structure with surviving timbers found at Killerby Quarry', https://www.archaeology.co.uk, 2019.

9 Zwerger, K., *Wood and Wood Joints: Building Traditions of Europe*, Japan and China, Birkhäuser, Switzerland, 2015.

10 Hebel, D.E., Heisel, F., Goidea, A., Floudas, D. and Andreen, D., 'Pulp Faction 3D Printed Material Assemblies though Microbial Biotransformation', in J. Burry, J. Sabin, B. Sheil and M. Skavara (eds), *Fabricate 2020*, UCL Press, London, 2020, pp. 42–9. Benjamin, D., 'Living Matter', in S. Tibbits (ed.), Active Matter, MIT Press, Boston, 2017, pp. 255–9.

11 Wurm, J., 'Bio-tech soundproofing: Optimising room acoustics with fungus-grown panels', https://www.arup.com/projects/biocomposite-systems-for-interior-fit-outs

12 Svilans, T., Tamke, M., Thomsen, M.R., Runberger, J., Strehlke, K. and Antemann, M., 'New Workflows for Digital Timber', in F. Bianconi and M. Filippucci (eds), *Digital Wood Design: Innovative Techniques of Representation in Architectural Design*, Springer International Publishing, Switzerland, 2019, pp. 93–134. Menges, A., Schwinn, T. and Krieg, O.D., 'Advancing Wood Architecture: An Introduction' in A. Menges, T. Schwinn and O.D. Krieg (eds), *Advancing Wood Architecture: A Computational Approach*, Routledge, New York, 2016, pp. 1–9.

13 Faircloth, B., *Plastics Now: On Architecture's Relationship to a Continuously Emerging Material*, Routledge, London, 2015.

14 Onen Cinar, S., Chong, Z.K., Kucuker, M.A., Wieczorek, N., Cengiz, U. and Kuchta, K., 'Bioplastic Production from Microalgae: A Review', *International Journal of Environmental Research and Public Health*, vol. 17., 2020.

15 Brigham, C., 'Biopolymers: Biodegradable Alternatives to Traditional Plastics', in B. Török and T. Dransfield (eds), *Green Chemistry*, Elsevier, Amsterdam, 2018, pp. 753–70. Gökce Özdamar, E., Ates, M., et. al., 'Architectural Vantage Point to Bioplastics in the Circular Economy', *Journal of Architectural Research and Development*, vol. 2, 2018, pp. 2–9.

16 Saba, N., Jawaid, M., Sultan, M.T.H. and Alothman, O.Y., 'Green Biocomposites for Structural Applications', in M. Jawaid, M.S. Salit and O.Y. Alothman (eds), *Green Biocomposites: Design and Applications*, Springer International Publishing, Switzerland, 2017, pp. 1–27. Fan, M. and Fu, F., *Advanced High Strength Natural Fibre Composites in Construction*, Woodhead Publishing, 2016. Dahy, H. and Knippers, J., 'Biopolymers and Biocomposites Based on Agricultural Residues' in Hebel, D.E. and Heisel, F. (eds), *Cultivated Building Materials*, Birkhauser, Berlin, 2017, pp. 116–23.

17 Sanandiya, N.D., Vijay, Y., Dimopoulou, M., Dritsas, S. and Fernandez, J.G., 'Large-scale additive manufacturing with bioinspired cellulosic materials', *Scientific Reports*, vol. 8, 2018, article number 8642.

18 Bryan, F.R., *Beyond the Model T: The Other Ventures of Henry Ford*, Wayne State University Press, Detroit, MI, 1997.

19 Dritsas, S., Vijay, Y., Dimopoulou, M., Sanadiya, N. and Fernandez, J.G., 'An Additive and Subtractive Process for Manufacturing with Natural Composites', in *Robotic Fabrication in Architecture*, Art and Design, Springer International Publishing, Switzerland, 2019, pp. 181–91.

20 Dade-Robertson, M., Rodriguez-Corral, J., Guyet, A., Mitrani, H., Wipat, A. and Zhang, M., 'Synthetic Biological Construction: Beyond "bio-inspired" in the design of new materials and fabrication systems', 3rd International Conference Biodigital: Architecture & Genetics, Barcelona, Newcastle University, 2017, pp. 292–301. Cogdell, C., *Toward a Living Architecture?: Complexism and Biology in Generative Design*, University of Minnesota Press, Minneapolis, 2019.

21 Armstrong, R., Ieropoulos, I., Wallis, L., You, J. and Nogales, J., 'Living Architecture: Metabolic applications for next-generation, selectively-programmable bioreactors', https://livingarchitecture-h2020.eu/2018/11/02/living-architecture-metabolic-applications-for-next-generation-selectively-programmable-bioreactors, 2018.

22 Cruz, M., Beckett, R. and Ruiz, J., 'Bioreceptive Concrete Wall', in M-A. Brayer, O. Zeitoun, S. Papapetros, S. Bianchini, E. Quinz and C. Collet (eds), MUTATIONS/CRÉATIONS: LA FABRIQUE DU VIVANT, Editions HYX, Centre Pompidou, 2019, chapter 7.

23 Wurm, J. and Pauli, M., 'SolarLeaf: The world's first bioreactive façade', *Arq: Architectural Research Quarterly*, vol. 20, 2016, pp. 73–9. Peruccio, P.P. and Vrenna, M., 'Design and microalgae. Sustainable systems for cities', *AGATHÓN | International Journal of Architecture*, Art and Design, vol. 6, 2019, pp. 218–27.

24 Landecker, H., *Culturing Life: How Cells Became Technologies*, vol. 24, Harvard University Press, Cambridge MA, 2010.

Stefana Parascho

Full-scale prototype of the project Cooperative Robotic Assembly. Over 300 elements were placed cooperatively by two robotic arms.

Home Position:

Reflecting on Disciplines, Discontinuities and Design Spaces

Home position is a robot's pre-defined configuration to which it returns after it has fulfilled its daily tasks. The home position never fails. It is always known, reachable, considered safe and will never throw an error. However, heading back to home position is also a retreating action that does not consider the movement along the way. The goal is the pose, with the path there being secondary.

Introduction

While architecture has ventured far outside its boundaries into more and more experimental technologies in the last decades, it has always remained connected to its own home position or comfort zone. Unlocking new technologies has meant leaving the field and reaching out to neighbouring disciplines. The difficulty has been bringing those back into the field so that they can be used, further developed and adjusted to serve the needs of the architectural discipline. The complexity of other fields' content and the different languages that are used has made it challenging to provide accessible tools and knowledge to architects.

  The construction sector – and the affiliated field of architecture – has historically lagged years behind other industries in adopting digital technologies. The reasons for this are multi-faceted, ranging from the challenges of individual building designs precluding their mass production to the time-consuming process of introducing new technologies into an inert domain with a strong sense of tradition. This means that architects have been on stand-by, watching other fields develop their own tools and technologies and waiting to benefit from their advancements and breakthroughs. By the time architects familiarised themselves with these new processes, they were bound to industry-made, pre-defined black-box tools, targeted at narrow and specific applications. These tools never had a critical impact on the field of architecture, but rather focused on industrial needs. What is missing are processes, tools and developments that:

· enable the architectural discipline to push its own boundaries,
· expand its reach through new design spaces, and
· impact the built environment, with all that it contains.

We are currently witnessing a shift in perspective. The fields of robotics, computer science and material engineering have entered architectural research and many architects have embraced new developments with unprecedented openness. However, there is an unresolved need for a robust and efficient accessibility channel between architecture and technical disciplines. Without it, designers are restricted to what other fields offer but have no say in what they really need. It does not suffice to take what has been developed and apply it to a different setting. We hit that ceiling years ago, with more and more robotic construction processes being implemented that do not substantially push architectural design forward (yet another robot process). What the discipline needs is to utilise the low-level technical knowledge, by giving it the space, time and acknowledgement for its value in a creative field.

On the other hand, the traditional design process is based on intuition, which is centred on working with well-known constraints and parameters. This process needs to adapt to create the space for new technical knowledge. We want to design architectural structures with complex fabrication processes in mind, that yield structurally efficient results and at the same time reinvent the design language. Our lack of intuitive understanding of digital methods, with their complexity exceeding our imagination, hinders us from designing with all their potentials in mind. If the design is considering the entire process, from material to fabrication, structure, function and formal expression, rather than assuming that the first four are already embedded into our intuition, we can allow for new developments in these areas to visibly impact design results. This means that we can fully explore a new design space uncovered by new technologies only through analysing, calculating and visualising their design possibilities.

Integrative design spaces

One means of tapping into the vast design space of robotic construction is by looking at multi-robotic processes. Pushing beyond the applications of single robotic arms, cooperative robotics allow us to take advantage of the capacity of performing multiple robotic tasks simultaneously. While it may seem straightforward to control a single industrial robot, employing more than one robot scales up the complexity of the task in a non-linear manner. When using one machine, we typically assign it a pre-defined task. In contrast, multiple robots can vary their tasks throughout the construction process, work independently or support each other. This gives space to new fabrication approaches and new roles for robots in construction. At the same time, it requires a more low-level understanding of a robot's functionality and control techniques. With the complexity of the robotic procedures also rises the complexity of the constraints resulting from a multi-robotic set-up. Adding multiple machines to the set-up results in multiple co-dependent constraints. These include:

· collisions between the different robots and between the robots and the built structure
· reachability constraints resulting from the distinct robot reach volumes and their intersection, and
· the constant changes in the environment in which the robots operate.

This means that the dynamics of the system in which the fabrication task is executed are not solely related to the robot itself but are strongly impacted by how it interacts with the other robots, the built structure and its environment.

If things get so complicated, why would we even want to use multiple machines? Multiple machines open up a vast number of options that are unimaginable using a single one. The processes made possible by the interaction of machines can achieve far more than the sum of their individual capabilities. Similar to humans, we can let the robots play different roles, such as feeding material, processing it, passing on elements, serving as supports, deforming or deconstructing elements, or even sensing, capturing and monitoring construction processes. And they push architects to explore the full potential of such machines, beyond applying a pre-existing routine with limited impacts on the design outcome.

The project Cooperative Robotic Assembly,[1] developed by the author at ETH Zurich, Gramazio Kohler Research, between 2014 and 2019, explores precisely the potential of such multi-robotic procedures. It proposes an assembly method for the scaffold-free

Architects have been on stand-by, watching other fields develop their own tools and technologies and waiting to benefit from their advancements and breakthroughs.

Cooperative Robotic Assembly process. The robots alternate between placing and supporting the structure.

construction of complex space-frame-like structures by alternating the placement of elements, while using the other robot as a support for the structure, whenever it is not actively placing an element.

The developed design system is based on the fabrication logic. It entails a new connection system, in which only two bars are connected to one another. The overall geometry is based on the aggregation of simplified tetrahedral assemblies, consisting of three bars per step so that the overall geometry can be expanded into all directions equally. The result is a space-filling truss system, which ensures static determinacy at every step and enables the robots to add one element at a time without compromising the stability of the entire structure. At the same time, it allows for differentiation regarding number, length, orientation and density of elements.

What is evident is that no part of the process can be subordinated to the others if the intention is to take full advantage of the potential of the cooperative robotic process. Geometry, material, structural behaviour,

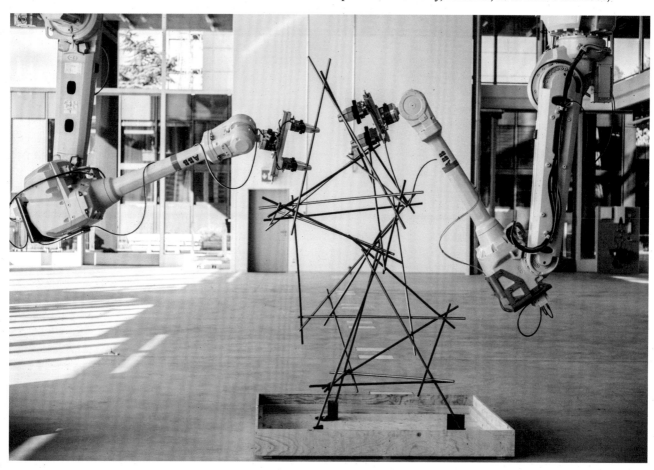

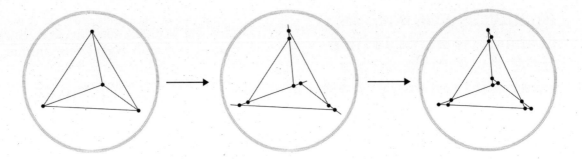

Evolution of connection system: (a) multiple bars connecting in one point, (b) bars connected to one existing bar, and (c) bars connecting to two existing bars for additional stiffness.

connection system, sequence, path-planning and fabrication procedures were developed in an integrated process where a decision is circled back into the overall design system as soon as it is being developed. As such, the design of the structure was naturally a result of all these parameters and methods, and could not have been developed independently.

The different building blocks of the project were developed in a rather organic manner, calibrating each of them regularly to match the other ones. From a robotics perspective, the sequence of instructions is straightforward to implement (except for the path-planning process required to avoid collisions during movements). However, from an architectural perspective, the cooperative robotic process paradoxically increases both the possibilities and the limitations for design. While the proposed method enables the construction of differentiated spatial structures, with individual dimensions and numbers of elements meeting at a node, it also limits the possible geometries to those that fulfil the constraints of the arrangement logic and enable collision-free robotic movements. In fact, a snowball-type of dependency evolved throughout the project: the connection system between metal bars was adjusted to enable the assembly of bars at individual angles without additional, customised connection elements. This led to a reduction of the stiffness of the structure since shifts in the connections introduced bending moments. To improve the structural performance, a second connection point to the existing structure was added per element, which led to a more constrained geometric system – each newly added bar now had to be tangent to two existing bars. Coupled with the overall aggregation logic that requires the placement of three new bars to form a tetrahedral assembly at every step, it resulted in a distinct solution space.

Even though the global structure allows for many solutions per new bar, and the number increases exponentially with the size of the structure to millions of

options, these are all discrete solutions. We can imagine the design space as a collection of points on a surface. When reversing the design logic and starting with a given structure – which would then be transformed into the proposed system – the chances are very high that an intuitively proposed structure would lie somewhere between the points of the actual design space, meaning it is not feasible with the given system.

So how do we engage with such a seemingly constrained design space? The problem is not the size of the space or its dimensionality. Even if solutions are like points on a canvas, their number is still far greater than that of a regular, standard space-frame. There is no limitation to the number of bars that can connect at one node or to the angle of the connection, and the bars are solely constrained to a minimum length by the size of the gripper. The problem is, rather, the unintuitive topology of the design space. We simply cannot imagine what a feasible solution is. Furthermore, there are gaps between those solutions, meaning if we find one that is feasible, we do not know which way to look for another solution.

This is where computational design tools come into play. No pre-existing tool will help understand this solution space. The difficulty of our discipline is that architects are often seen as end-users who should not spend their time on developing tools but be handed over these tools by specialists. It lowers the ceiling, limiting design possibilities, no matter how many tools other fields provide.

In the case of the presented project, the chosen approach was to develop a custom computational tool and hand over the design to an exploratory optimisation process which computed the solution space resulting from the geometric/structural and fabrication-informed constraints.[2] This allowed for feasible solutions to be generated at the expense of outside control. However, it forced us to consider every dependency within the system as a whole and translate it into instructions that allowed for an optimisation process to take place

Intelligent Control: Disruptive Technologies

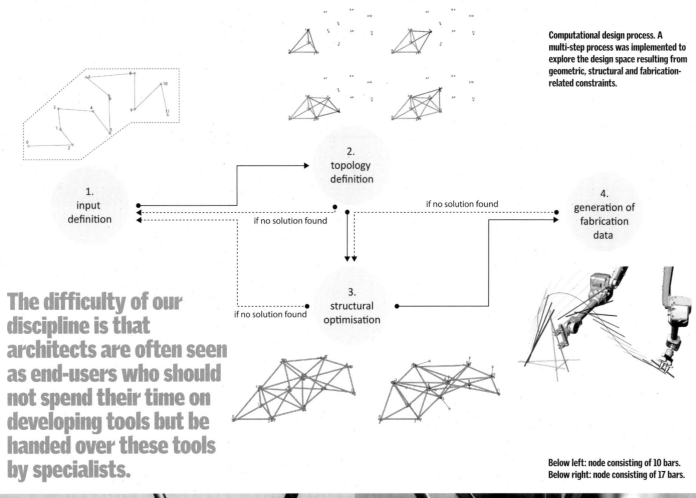

Computational design process. A multi-step process was implemented to explore the design space resulting from geometric, structural and fabrication-related constraints.

1.
input
definition

2.
topology
definition

if no solution found

4.
generation of
fabrication
data

if no solution found

3.
structural
optimisation

if no solution found

The difficulty of our discipline is that architects are often seen as end-users who should not spend their time on developing tools but be handed over these tools by specialists.

Below left: node consisting of 10 bars.
Below right: node consisting of 17 bars.

Visualising design spaces

An alternative approach to working with unpredictable design spaces is visualising the feasible space through various media. During the architecture master's degree course taught at Princeton University, Computational Fabrication in Architecture, students are tasked with developing a robotic process based on their ideas of possible applications. The process design exercise is preceded by a thorough introduction to algorithmic thinking, coding, robotics basics and a custom robotic control tool. Ultimately, students need to combine this new knowledge into a functional prototypical fabrication process. Instead of relying on existing tools that would allow them to execute pre-defined movements, they are exposed to the low-level specifics of robotic control, kinematics, tool-design, coordinate transformations and set-up design. While sometimes an agonising experience, the goal is to enable them to understand and visualise the possibilities of their chosen robotic system and design processes that overcome the limits of pre-existing robotic processes popular in architecture schools, such as drawing, pick-and-place or wire-cutting.

The project, A Book of Folds, by students Chase Galis and Chris Myefski, investigated the potential to use robots for paper-folding. Unlike existing folding processes in which a sheet element is carved in a first step and in a second step folded by hand or robots,[3] the proposed process involved executing the folding in one automated sequence. This meant that the design of the robot end-effector, the investigation of possible folding motions, the exploration of the robot's reach and, in particular, the understanding of the sequence of folding instructions and its implications were crucial to identifying feasible robotic folding patterns. While the project is not targeting a novel material fabrication method, it is particularly successful in exposing the capabilities of the robot beyond the well-known additive and subtractive processes. The success of the project lies in the combination of systematic material tests and thorough tests of the robotic procedure, leading to a physical catalogue of possible results. It served as a literal visualisation of the design space and enabled the students to imagine the potentials opened up by this specific process. The visualisation does not have to imply a digital 3D model of all possible solutions (which is often challenging to generate and read). Instead, it can be represented, as in this case, by a collection of physical results which map the specific solution space of the process.

Using a similar methodology, the project Throwing Clay, by Catherine Ahn, Piao Liu and Lisa

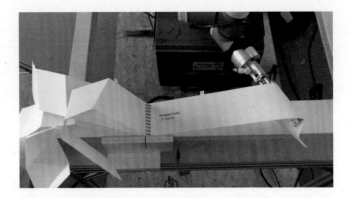

A Book of Folds. Robotic folding process. Project and image by Chase Galis and Chris Myefski.

A Book of Folds. Before and after folding process. Project and images by Chase Galis and Chris Myefski.

Ramsburg, investigates the potential of robots throwing objects. Besides identifying a suitable tool, movements and speed required for a controlled throwing procedure, the project tackles the stochastic element of this process. Even with the same settings, tool and object, the final position of the thrown object varies within a certain range. The purpose of the project is again not to implement a specific manufacturing process through robotic fabrication but to systematically investigate the components that contribute to the definition of the solution space resulting from the robotic movement. The investigations ranged from analysing the human throwing motion and mapping it on to a robotic arm, exploring several end-effectors based on existing throwing and catapulting tools, and analysing different objects and their behaviour when thrown by the robotic arm. Ultimately, the design space was mapped through a combination of visual registration of the results and pre-computed possibilities based on the robot arm's kinematics.

What both projects show is the importance of shifting the expectations of student projects from quickly replicating a complex process through existing tools to learning the underlying concepts, methods and techniques of robotic fabrication and empowering students to imagine new implementations. Breaking down a robotic process into a series of steps – from determining the robotic set-up, designing and testing an end-effector, and testing the material behaviour, up to defining and testing the robotic movements via low-level commands – enables an expansion of the understanding and imagination of the capabilities of such tools, and opens up new perspectives. Part of this process is understanding that while there are tools readily available to simplify the execution process, using these without engaging with the underlying knowledge will limit their capability of accessing the totality of the design space uncovered by robotic techniques.

While architecture is opening up towards other fields and new technologies, there is still a need for widening our perspective to include the design process besides the design result.

Bottom left: Throwing Clay. Registration of throwing results on a clay base serving as a visualisation of the design space. Project and image by Catherine Ahn, Piao Liu and Lisa Ramsburg.

Below: Throwing Clay. Mapping of the robotic movement and axis rotations. Project and image by Catherine Ahn, Piao Liu and Lisa Ramsburg.

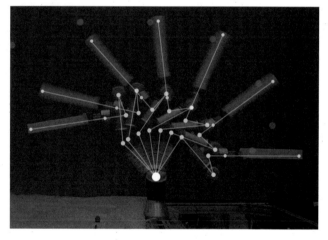

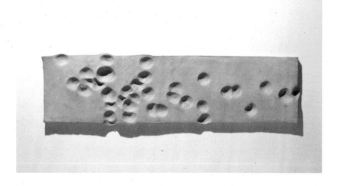

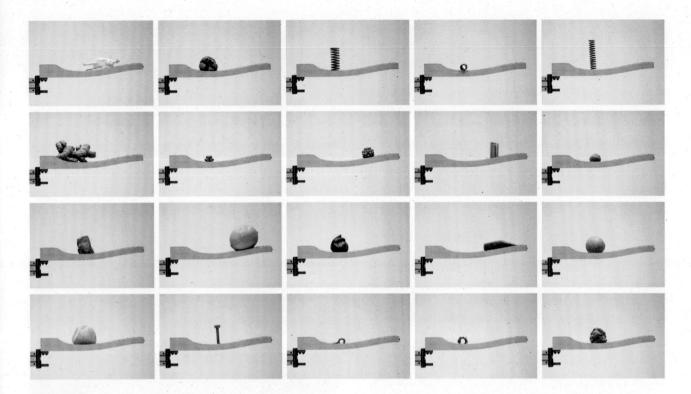

Conclusion

While architecture is opening up towards other fields and new technologies, there is still a need for widening our perspective to include the design process besides the design result. Ready-made tools that allow designers to apply a fabrication process without questioning its possibilities are more limiting than liberating and are leading to uniform results, far away from the idea of innovation. At the same time, the technical knowledge required to innovate would quickly exceed the possibilities of design education and would turn every project involving a robotic process into a years-long research endeavour. What the discipline needs is bridges to neighbouring fields, based on a common language used to clearly communicate the technical concepts and methods to designers and architects. We need tools that help us to understand and build an intuition for these technologies through visualisation, analysis and feedback, instead of tools that substitute the implementation process by skipping this phase. Similarly, students benefit from learning to engage with new technologies at all levels and understanding how to exploit their potential rather than how to execute pre-existing processes. By allowing a project to be purely explorative, to uncover the design space of a novel robotic process, we learn to engage with a different aspect of design. Systematically recording, analysing, tracing and translating processes into coding procedures are all ways of connecting to other scientific fields. Providing an environment for this in design education represents one step towards achieving a tighter connection to these disciplines.

Instead of focusing on implementing a specific fabrication process, projects can include an explorative phase to unveil what possibilities and limitations such a procedure brings. Ultimately, what matters is what these new tools and knowledge can bring to the table in terms of design possibilities, and how they can push forward our discipline beyond the fascinating complexity of robots and algorithms. Once we uncover this, we can return to our home position with new skills and knowledge to advance our field.

The blurred borders that we already see between architecture and technical disciplines, as well as the increasing number of cross-discipline collaborations that many projects in the field are built upon, are an indication of a changing discipline. We are working towards a future in which the architect is no longer the representation of the individual creative genius but part of an inclusive team of practitioners or researchers that will enable us to fully engage with new advancements inside and outside the field.

Throwing Clay. Testing
and recording of different
thrown objects. Project
and image by Catherine
Ahn, Piao Liu and Lisa
Ramsburg.

What the discipline needs is
bridges to neighbouring fields,
based on a common language
used to clearly communicate the
technical concepts and methods
to designers and architects.

Acknowledgements
Cooperative Robotic Assembly was developed under the supervision of Prof. Fabio
Gramazio, Prof. Matthias Kohler and Prof. Dr Stelian Coros at ETH Zurich. The presented
student projects were developed in the course Computational Fabrication at Princeton
University: A Book of Folds by Chase Galis and Chris Myefski and Throwing Clay by
Catherine Ahn, Piao Liu and Lisa Ramsburg.

1   Parascho, S., 'Cooperative Robotic Assembly: Computational Design and Robotic
    Fabrication of Spatial Metal Structures', unpublished doctoral thesis, ETH Zurich, 2019,
    https://doi.org/10.3929/ethz-b-000364322
2   Parascho, S. et. al., 'Computational Design of Robotically Assembled Spatial Structures:
    A Sequence Based Method for the Generation and Evaluation of Structures Fabricated
    with Cooperating Robots', in *AAG 2018: Advances in Architectural Geometry*, 2018,
    presented at the Advances in Architectural Geometry, Klein Publishing, 2018, pp.
    112–39, https://www.research-collection.ethz.ch/handle/20.500.11850/298876, accessed
    30 August 2020.
3   'RoboFold', https://www.robofold.com/make/technology, accessed 30 August 2020.

Automation, Architecture and Labour

Mollie Claypool

Increased automation over the last decade has transformed much of the way architects and designers interact, communicate and produce work. Platforms – connecting us through the power of computation and the Internet of Things – and the economic model for automation, have proliferated into our daily lives. We are reliant on these platforms. When the internet goes down or when a software app will not update, our capacity to participate in the economy, to produce and reproduce ceases.

House prices in the UK have increased up to 120% in the last decade,[1] and affordability continues to worsen.[2] In England alone, over 8.4 million (one in seven) people are affected.[3] Yet state capital spent on housing has decreased by 50% in the last 20 years.[4] Lack of vision, minimal investment and deregulation by the state since the 1970s has left the sector extremely susceptible to economic actors that see an opportunity to make a profit. And this is not a local condition. Worldwide, over two billion homes need to be built in the better part of the next century.[5]

The architecture and construction industries are also not well situated to deliver on these needs. Architecture is highly marginalised and 'there is almost no convincing evidence of the value of architects'.[6] Construction is a highly striated industry that suffers from a significant labour shortage,[7] is one of the least digitised industries worldwide (second only to hunting)[8] and productivity has flatlined since the middle of the twentieth century.[9]

This sits in stark contrast to other industries that have been able to adopt to (or emerge due to) digitisation, where growth and productivity has skyrocketed. And an increasing wealth divide has emerged. Capital is more and more centralised with the richest 1% owning 44% of the world's wealth.[10] It is not a

INT: Robotic Building Blocks. Half-chair, chair, column. By Zoey Tan, Claudia Tanskanen, Qianyi Li and Xiaolin Yin, RC4, Design Computation Lab, The Bartlett School of Architecture, UCL, 2016.

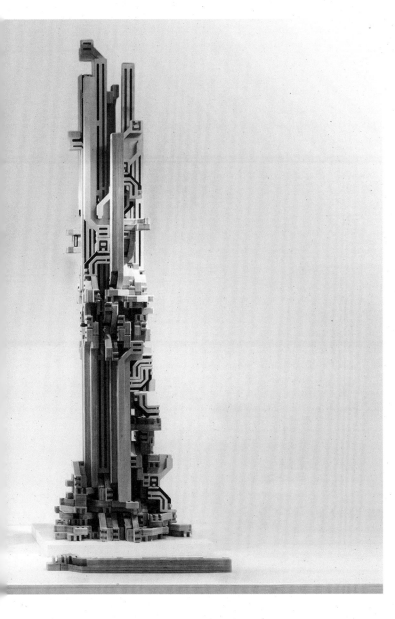

coincidence that this divide runs in parallel to increasing automation in most industries.

And so it is this intersection between the distribution of automation, capital, activism, housing and people that the work of Automated Architecture Ltd (AUAR) and research laboratory AUAR Labs at The Bartlett School of Architecture, University College London (UCL) is situated. The following article will provide the broader context for the work of AUAR and design research of AUAR Labs. Originally grounded in experimental design speculations by our students in both MArch Unit 19 in the MArch Architecture programme from 2012–18 and Research Cluster 4 (RC4) in the post-graduate MArch Architectural Design programme from 2015 onwards, the labour of our AUAR/AUAR Labs has shifted this work outside the realm of academia into working directly with the communities most affected by the increasing financialisation of housing and automation.

Automation as a design project

For decades, the dominant narrative around automation in construction has centred on the loss of manual labour jobs. Often the digital fabrication technologies that are developed to replace these jobs focus purely on the productivity gained by implementing the use of a single-task robot – such as a bricklaying robot,[11] a tile-laying robot or a welding robot – and increased productivity typically increases the wealth for the owners of these technologies. Unfortunately, these robots require significant investment only possible with the backing of venture capital and are often only available to those with extreme wealth, as evidenced by Construction Robotics' bricklaying robot which cost US$500,000.[12] Yet, while they are technical innovations in digital fabrication or digital construction, they are often extremely limited in solving narrow problems.

Investment in construction automation has also occurred at the scale of the factory. In 2019 Katerra, a California-based company creating modular building systems, primarily for housing, became the first unicorn company in construction, taking in over US$1 billion in investment. Taking a similar approach to automation as automobile manufacturing, modular housing factories such as Katerra resulted in the centralisation of housing production to off-site factories, disconnected to the contexts in which these systems are deployed. Furthermore, the workers that build these housing systems are currently imported on to sites from elsewhere, resulting in the marginalisation of local employment and economic development opportunities. Factory-made housing is seen as displacing local jobs, which creates resistance and scepticism among communities and creates limited localised capacity for housing produced using automation.[13]

A critique of the single-task robot, as well as the centralisation of automation that disables access to technology by local communities and contexts, can be connected to neoliberal capitalism in architectural production. Automation only in the terms of digital design and digital fabrication are what Nick Srnicek and Alex Williams refer to as 'folk politics'. These are 'tactics and strategies which were previously capable of transforming collective power into emancipatory gains' but are now so divorced from actual mechanisms of power that they are 'incapable of transforming capitalism' and have become 'drained of their effectiveness'.[14] Digital design and digital fabrication – known collectively as 'digital X' – have been long heralded within the discipline as being transformative.

Yet, by being focused on solving small-scale problems within the discipline, digital design (focused on representation, variation and affect) and digital fabrication (focused on replicating craft) effectively enable collective disempowerment and marginalisation of architects within the discourse on automation. Other actors, such as those with capital to develop and implement automation, and therefore economically benefit from it, have emerged as a result of the continued acceptance of the transformative promise of the digital made almost 30 years ago. This despite evidence pointing in the opposite direction towards the failure of the digital to radically democratise production.

The way things are done within architecture is not just accidental. Disciplinary practices are historically constructed.[15] In the last several decades, these disciplinary practices benefit neoliberalism, which suppresses the strength and power of the local or small scale, effectively dismantling the power of 'bottom-up' change (hence resulting in folk politics). As architects have become more marginalised, they too have become small-scale actors, and thus increasingly suppressed. To broaden the scope of automation in architecture from the discourse around the digital X or historical disciplinary practices and towards automation is emancipatory, both for the practice of architecture as well as the people that it should be serving. Architecture in the age of automation can no longer ignore that it too must change.

Design is politics. Design operates not just at the scale of the technological: it is social, it is economic.

Architects can either contribute to changing notions of the role and value of design or succumb to greater irrelevance. Core to the work of AUAR/AUAR Labs is the notion that automation is a design project.[16] When framed as a design project, automation becomes an arena through which architects can raise and discuss issues such as ownership, distribution, labour and the culture and impact of automation on architectural production. These are shared issues and global issues that transcend place, cultures and contexts. By addressing them through automation, architects can understand the design of automation in architectural production as a collective social project.

Architects can also design a collective division for automation. Perhaps, as the political theory of accelerationism argues, capitalism can be repurposed, by 'preserv[ing] the gains of late capitalism while going further than its value system, governance structures, and mass pathologies will allow'.[17] Therefore, this vision can be one that does not centralise or monopolise automation, as neoliberalism does, but demonstrates the capacity for automation to accelerate the production of distributed abundance and amplify interdependence,[18] instead of enabling further austerity, divisiveness and marginalisation. A vision for automation in architecture must:

· Confront the value systems and hierarchies embedded in both architecture's social and spatial practices by radically rethinking architectural syntax in terms of geometry, tectonics and aesthetics.
· Redefine automation's capacity to enable increasing access to architectural production for the benefit of the many, rather than the few.
· Consider automation as multi-scalar – as both 'on the ground' tech for design production and as a framework for coordination and logistics across scales and contexts.

Architecture for automation

Fulfilling this vision requires designing architecture for automation. To do so is to go to the core of architecture, which for thousands of years has been designed in relationship to localised contexts and resources. Yet increasing globalisation since the Second World War has disrupted this tradition: a building can be designed in New York with parts manufactured in Beijing, Germany or Shanghai and be built in Cairo, Hong Kong or Johannesburg. It is now evident that this practise is unsustainable, with the building and construction industry contributing to 39% of all carbon emissions.[19]

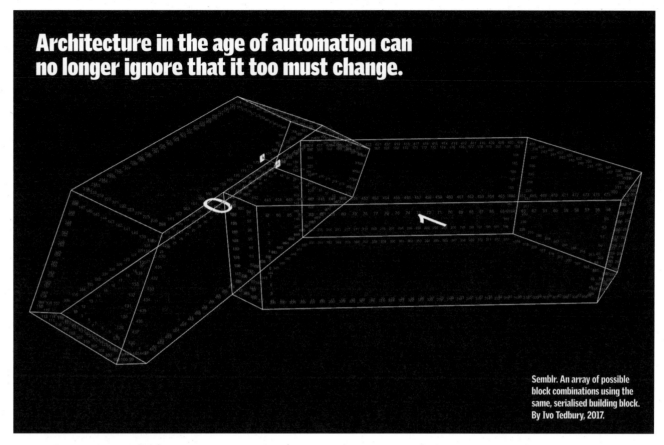

Architecture in the age of automation can no longer ignore that it too must change.

Semblr. An array of possible block combinations using the same, serialised building block. By Ivo Tedbury, 2017.

Thirty years of digital design has further contributed to this, enabling architects to design complex and intricate geometries. These buildings have thousands of bespoke, one-off parts costly to manufacture in a highly striated and inefficient production chain. It is not a kind of architecture that can speak to automation, which requires similarity, repetition and seriality. A new generation of digital architects and designers have confronted this contradiction by radically rethinking the parts that make up buildings utilising the concept of the Discrete.[20]

The Discrete builds on the notion of digital materials, or a new kind of building block 'assembled from a discrete set of parts, reversibly joined in a discrete set of relative positions and orientations' that has the same structure as data in a computer program. These new, wholly digital building blocks can then be organised into different positions, which in principle can be continuously altered.[21] This dramatically reduces the number of different kinds of parts that make up a building. The same or similar building blocks are able to create part-to-whole[22] relationships that enables patterns[23] or assemblies to emerge from the building

blocks being combined to serve different architectural functions. Staircases, beams, slabs and columns can all be designed using a single block.

A Discrete architectural syntax can act as a universal framework for architectural production, supporting diversity of geometries, forms and tectonics to emerge in response to particular contexts. It is also well-suited for increasing automation. 3D printing, robotic assembly or other digital fabrication methods, such as CNC machines, can be used by a design project employing this thinking. The Discrete shortens the production chain for buildings without increasing costs, as the same part can be used to form the entirety of a building's structure and spatial configurations.

It is within this context that projects such as Semblr by Ivo Tedbury (2017) and Automated Living System (ALIS) by Akhmet Khakimov, Estefania Barrios, Evgenia Krassakopoulou, Joana Correia and Kevin Saey (2019), student work in Unit 19 and RC4, have been developed. Each of the projects developed a single, repeatable building block that could be manufactured using a common digital fabrication technology like a CNC machine. It also presents the challenge of

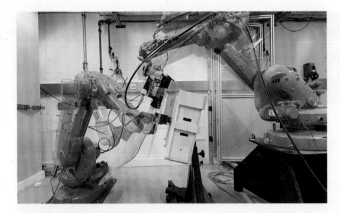

Semblr. Robotic assembly of a small house using one single repeated building block and distributed self-similar modular robots. By Ivo Tedbury, 2017.

Pizzabots. Distributed modular Pizzabot robots moving active building elements for the assembly of a small house. By Mengyu Huang, Dafni Katrkakalidi, Martha Masli, Man Nguyen and Wenji Zhang, RC4, Design Computation Lab, The Bartlett School of Architecture, UCL, 2018.

developing a Discrete building block that creates minimal waste. Each project uses only a single sheet of plywood to form one block, sitting in direct contrast to earlier 'digital' work where variation in size of parts results in significant material waste.

Automation for assembly

Part-to-whole relations in a Discrete framework also enables automation to be used at the scale of the building assembly, distributing the labour it takes to construct a building away from the manual labour of construction workers and into robotic automation, which has previously been called Discrete Automation.[24] It throws into question the notion of ownership of space: could housing be shared, used and inhabited in new ways due to automation? In ALIS, Semblr, Pizzabots by Mengyu Huang, Dafni Katrkakalidi, Martha Masli, Man Nguyen and Wenji Zhang (2019) and MOBO by Po-Fu Yang, Nadia Saki, MengMeng Zhao, Tian Chuan and Keshia Lim (RC4, 2019), the potential of modular robotic assembly was explored to begin to answer this question.

Tedbury developed a distributed modular robot capable of distributing the building blocks, enabling homes to be created quickly with minimal human

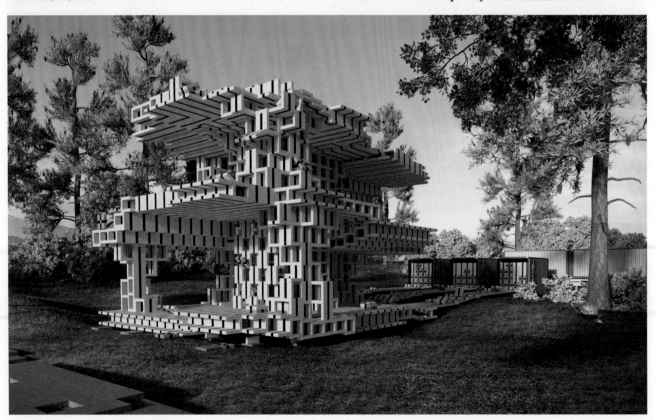

intervention. He has extended this work post-graduation from Unit 19 in the start-up company Semblr, further developing it into novel modular bricklaying robots.[25] The distributed modular robots developed for ALIS more closely mimicked the geometry of the building block itself. Spaces unused in an apartment building during a typical workday were 'activated' by these robots, with the building able to change and adapt throughout the day to changing needs of its inhabitants, determined via a mobile app. This allowed all space available within a building to be used – whether it was being used for work or rest – through ongoing robotic assembly and disassembly of different spatial configurations. Pizzabots almost entirely blurred the boundaries between what was 'building' and what was 'automation', designing a robot the exact geometry of the building block itself (the size of a pizza box), enabling it to entirely merge with and negotiate the configuration and reconfiguration of the building. In MOBO, the robot was made of modular parts that each carry out a single type of movement and can be combined in different ways, resulting in an ecology of construction robots fit for a variety of different assembly tasks.

The work sits in direct relationship to the work of other researchers such as Maria Yablonina, who envisions small robots working alongside workers that are 'continuously performing construction and spatial reconfiguration tasks in response to their human co-habitants'.[26] This raises crucial questions around ownership, decision-making and data, and has significant ecological implications, as building blocks can be continuously re-used into other spatial assemblies. Yet these projects also present a more speculative approach to automation: we are not yet quite in a world where distributed modular robots will be on building sites or living and working amongst people in buildings, negotiated through an app. Our assumptions about what a home is sits in opposition to the temporality of Discrete Automation. So how can automation be implemented in architecture today?

MOBO robot and construction workers during building assembly of an apartment building. By Po-Fu Yang, Nadia Saki, MengMeng Zhao, Tian Chuan and Keshia Lim, RC4, Design Computation Lab, The Bartlett School of Architecture, UCL, 2019.

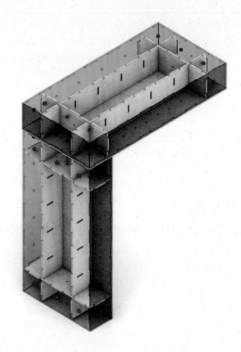

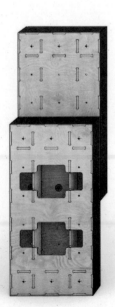

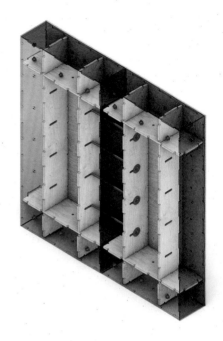

Our assumptions about what a home is sits in opposition to the temporality of Discrete Automation. So how can automation be implemented in architecture today?

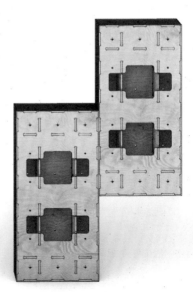

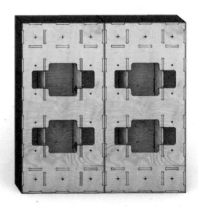

The possible connections of Block Type A using post-tensioning. Automated Architecture (AUAR), 2020.

An AUAR member training a workshop participant in how to use AUAR's web-based combinatorial design app. Automated Architecture (AUAR), 2020.

How can communities access automation without full automation? How can new understandings about home and place emerge? A possible answer lies in platforms: the economic model of automation.

Platforms as collective resilience

The distribution of automation into communities requires, as argued by Nick Srnicek in *Platform Capitalism* (2017), that the rise of platforms within capitalism (Facebook, AirBnB, Uber) are understood not as cultural actors but as extractive economic actors, out to make the most profit from citizens that use their platforms. As such, these companies are seen not as creating platforms for the common 'good' but in the pursuit of power, extraction and wealth.[27] Srnicek also predicts that due to the rising inequalities produced by these platforms, eventually platforms as a business model will fail. His solution is to suggest the notion of the 'public platform'.[28]

Students in Unit 19 and RC4 have explored the public platform through the development of mobile apps owned by cooperative housing companies for connecting people to housing continuously adapted by distributed robots in response to a community's changing needs. The work of AUAR/AUAR Labs has focused on the public platform at three scales:

1. A Discrete building system called Block Type A that anticipates increasing automation while reducing the threshold to access by local communities.
2. A co-designed and values-centred[29] software platform that enables people to access Block Type from design to assembly/reassembly.
3. Increasing the capacity for digital labour within local communities that are building their own homes.

A development of the CNC'd building blocks from ALIS, Block Type A is a Discrete building system for housing using a timber block that is post-tensioned locally to achieve global stability in larger assemblies. Through distribution of the skills, tools and technologies to make Block Type A into communities, the system sits in opposition to a centralised, off-site factory model. The design of the block significantly reduces

Block West at Knowle West
Media Centre, Bristol, UK.
Automated Architecture
(AUAR), 2020.

How can communities access automation without full automation? How can new understandings about home and place emerge? A possible answer lies in platforms: the economic model of automation.

the access threshold for community members, with minimal training and no specialist tools required for prefabrication or assembly. Block Type A, in particular, is well-suited for full robotic prefabrication and anticipates this as our projects build up technological infrastructure in the communities in which we are working.

AUAR has developed an ecology of apps for using Block Type A and other Discrete building systems under development. These tools form the base technologies of a platform, as they link together design, fabrication and assembly in accessible ways. Co-designed with local communities, as well as trades and craftspeople, the software tools have taken a values-centred approach: putting communities that want to use these tools at the centre of the tool development. The community member's expertise and lived experience directly informs the capacity of the apps, providing instructive, game-like environments that empower the community to design, fabricate and build the homes they need.

In the recent project, Block West (2020), designed over a six-month period in Knowle West, Bristol,

25 community members aged between 12 and 76 participated in the prefabrication and assembly of a housing and community space prototype using 145 Block Type A over a ten-day period. The project provided part-time jobs in the local community throughout the project, serving as a prototype for community engagement, localised investment and pilot digital construction skills training programme for AUAR Labs' partners Knowle West Media Centres' 'We Can Make Homes' programme. This is a community-led housing project ongoing in Knowle West that is about to break ground for their first two homes. This is important in such an area where there is no centre or community hub, and where there has been historically little investment. It is a place where money often washes through rather than adding sustainable value or social infrastructure.

Automation as new citizenship

The distribution of skills, knowledge and expertise about Discrete Automation within local communities has served as a first stage of platform development, where citizens are activated in new ways to take ownership of their community and built environment. Digital labour becomes about investing in local skills and knowledge,

Block West at Knowle West Media Centre, Bristol, UK. Automated Architecture (AUAR), 2020.

Digital labour becomes about investing in local skills and knowledge, levelling up communities to further scale and coordinate efforts across the community and beyond.

levelling up communities to further scale and coordinate efforts across the community and beyond. As a Block West participant said, 'We are literally building the community from the bottom up. [The prototype] I've helped make isn't mine. It's ours. That's the most important thing to come out of this – it's given me ownership of my community. It's giving people different choices, better choices about how things can be. And it feels like only the beginning.'

This work aims to create collective resilience within the communities most at risk of the 'status quo' of automation in housing production: materially, technologically and socially. As a process, projects like ALIS, Semblr and other student work serve as a catalyst for projects such as Block West, demonstrating the speculative vision for automation in housing. AUAR/ AUAR Labs' projects develop the real, local technological and social infrastructure for these visions to be eventually realised when automation becomes more distributed, affordable and thus more accessible. These projects, by serving as prototypes for AUAR/AUAR Labs, are able to be replicated and adapted into other localised contexts. It is a realised example of how automation can be multi-scalar and values-centred. This approach sits at the intersection of communities, research and practice. It also creates novel possibilities around the distribution of automation, capital and housing, enabling new forms of engaged and empowered citizenship to emerge within the communities in which we are working. This requires an activist-based approach: it is hard and long but rewarding work. To do this is to embed ourselves within the communities in which we are working, to gain and build trust and to break down disciplinary silos between academia and practice, between communities and experts, in order to enact the changes we hope to see in housing and automation.

1 Brignall, M., 'Where have UK house prices increased most – and least – since 2010', *The Guardian*, https://www.theguardian.com/money/2020/oct/10/where-have-uk-house-prices-increased-most-and-least-since-2010, 10 October 2020, accessed 10 October 2020.
2 'Housing affordability in England and Wales: 2019', Office for National Statistics, 2019.
3 One in seven people in England directly hit by the housing crisis, National Federation, https://www.housing.org.uk/news-and-blogs/news/1-in-7-people-in-england-directly-hit-by-the-housing-crisis/, 2019, accessed 20 September 2020.
4 'The story of social housing', Shelter, https://england.shelter.org.uk/support_us/campaigns/story_of_social_housing, accessed 10 September 2020.
5 S. Smith, 'The world needs to build 2 billion new homes over the next 80 years', World Economic Forum, https://www.weforum.org/agenda/2018/03/the-world-needs-to-build-more-than-two-billion-new-homes-over-the-next-80-years, 2 March 2018, accessed 20 September 2020.
6 Callway, R., Farrelly, L. and Samuel, F., 'The Value of Design and the Role of Architects: A study for ACE Architects' Council of Europe prepared by the School of Architecture University of Reading', https://www.ace-cae.eu/uploads/tx_jidocumentsview/Value_of_Design.pdf, March 2019, accessed 10 October 2020.
7 MacKenzie, S., Kilpatrick, A.R. and Akintoye, A., 'UK construction skills shortage response strategies and an analysis of industry perceptions', *Construction Management and Economics*, vol. 18, 2000, p. 852.
8 Agarwal, R., Chandrasekaran, S. and Sridhar, M., 'Imagining construction's digital future', McKinsey Global Institute, https://www.mckinsey.com/business-functions/operations/our-insights/imagining-constructions-digital-future, 24 June 2016, accessed 10 October 2020.
9 *Ibid.*
10 'Global Inequality', Inequality Org, https://inequality.org/facts/global-inequality/, accessed 8 August 2020.
11 'SAM 100', Construction Robotics, https://www.construction-robotics.com/sam100/, accessed 8 August 2019.
12 'Bricklaying robot SAM – on your site for £330,000', *Construction Manager*, https://www.constructionmanagermagazine.com/s1am-bri1ck-laying-rob2ot-price-tag/, 4 September 2015, accessed 8 August 2019.
13 See Mean, M., White, C. and Lasota, E., *We Can Make: Civic Innovation in Housing*, https://issuu.com/knowlewestmedia/docs/wecanmake, 2017.
14 Srnicek, N. and Williams, A., *Inventing the Future: Postcapitalism and a World Without Work*, Verso, London, 2015, p. 10.
15 *Ibid.*
16 Claypool, M., Garcia, M.J., Retsin, G. and Soler, V., *Robotic Building: Architecture in the Age of Automation*, Detail Edition, Munich, 2019.
17 Srnicek, N., and Williams, A., 'Accelerate Manifesto for an Accelerationist Politics', *Critical Legal Thinking*, http://criticallegalthinking.com/2013/05/14/accelerate-manifesto-for-an-accelerationist-politics/, 14 May 2013, accessed 9 September 2020.
18 This term can be credited to Amahra Spence, creative director of MAIA Group (Birmingham, UK), who used it to describe a core goal of MAIA's work at the event *Making Together: Exploring New Tech, Tools and Tactics to Level Up Communities* at Bristol Housing Festival, October 2020.
19 https://www.worldgbc.org/sites/default/files/WorldGBC_Bringing_Embodied_Carbon_Upfront.pdf, p. 7.
20 For an overview of the architects and designers working on this topic, see G. Retsin (ed.), *Architectural Design*, vol. 89, no. 2, Wiley & Sons, New Jersey, 2019.
21 Gershenfeld, N., Carney, M., Jenett, B., Calisch, S. and Wilson, S., 'Macrofabrication with Digital Materials: Robotic Assembly', *Architectural Design*, vol. 85, no. 5, Wiley & Sons, 2015.
22 Koehler, D., *The Mereological City: a reading of the works of Ludwig Hilberseimer*, transcript, 2016.
23 Sanchez, J., *Architecture for the Commons: Participatory Systems in the Age of Platforms*, Routledge, London, 2020, pp. 80–82.
24 Claypool, M., 'Discrete Automation', e-flux architecture, https://www.e-flux.com/architecture/becoming-digital/248060/discrete-automation/, 2019, accessed 1 August 2020.
25 https://semblr.tech.
26 SPACE10. https://space10.com/project/digital-in-architecture/.
27 Srnicek, N., *Platform Capitalism*, Polity Press, Cambridge, 2017.
28 *Ibid.*, p. 128.
29 Van der Velden, M. and Mortberg, C., 'Participatory Design and Design for Values', *Handbook for Ethics, Values and Technological Design*, Spring 2014, pp. 1–22.

Knowle West Community team during the build of Block West at Knowle West Media Centre, Bristol, UK. Automated Architecture (AUAR), 2020.

Augmenting Human Designers and Builders:

Augmentation Discussed in Architectural Design Research

Soomeen Hahm

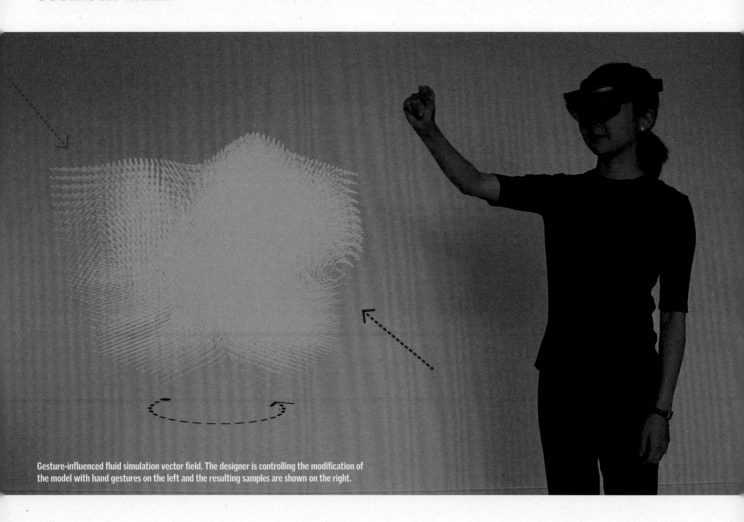

Gesture-influenced fluid simulation vector field. The designer is controlling the modification of the model with hand gestures on the left and the resulting samples are shown on the right.

For the past decade, the human role has often been neglected in discussions of computational design and automation. However, rapid developments in Artificial Intelligence (AI) and Augmented Reality (AR) are opening the possibilities for a closer relationship between humans, computers and machines, requiring multiple fields and industries to rethink the role of humans in the production chain. It raises an interesting question about how these developments will impact on the architectural design and construction industries, as well as how it will shape the physical environment of our future in a broader sense. In this article, several research projects are presented as case studies to discuss the issue and pose further questions on how architects should respond.

Following the internet age, it has been suggested that we are entering a new era of the 'Augmented Age'.[1] Physicist Michio Kaku imagines the future in which architects rely heavily on AR technologies.[2] Although it is not a new concept, the technology has only broken into the mainstream with the development of consumer AR devices[3] in the last five years. This has rapidly opened up possibilities in almost every aspect of our daily lives and is expected to greatly impact every field in the near future, including design and fabrication.[4] The timing is right to think about the impact of Mixed Reality (MR)/Augmented Reality (AR) technologies on the building industry.

With the commercialisation of various devices, Virtual Reality (VR) and AR are becoming increasingly popular topics in numerous industries, including design and architecture. Currently, within the architectural practices it is predominantly used for VR-/AR-integrated BIM management and presentation purposes. However, while being used as a powerful tool for the visualisation of building data and immersion into rendered spaces, the potential of MRs in the design process has remained largely unexplored.

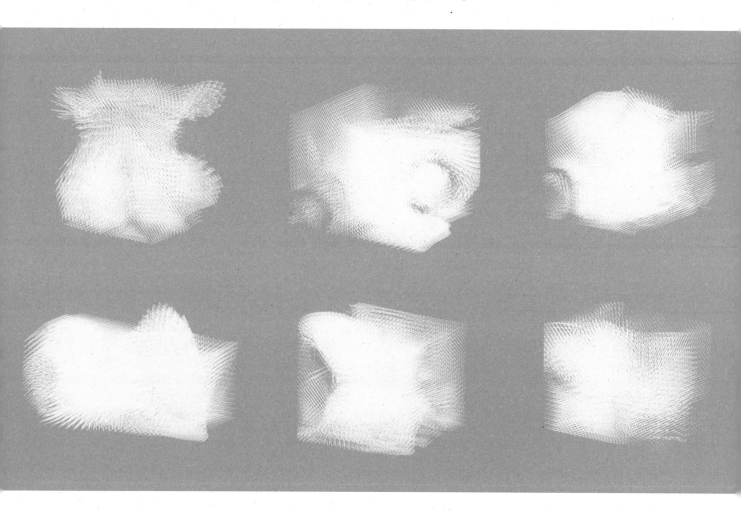

In academia, there has been a growing global trend of experimenting with the impact of VR/AR in space and city design. Many well-known academics have been working with the impact of personalised interactions on the modelling of cities. Similar projects dealing with psychologies of human experiences are widespread within the media arts. While most of these studies are confined to interactions with virtual matter and intangible worlds, there have been few studies examining the impact of these technologies on physical reality. For this reason, we directed our research towards the interface between real and virtual, and the dialogue of these two disconnected worlds through data.

It is important to note that 'augmentation' as a term covers a broad spectrum. It is a change of state through addition or extension. Augmentation can be applied to a physical reality, but it can also relate to the extension of human capabilities. This article focuses on the latter, posing questions on the future of design and construction, exemplified through various research projects. The following sections introduce a few projects that we have developed within the author's practice, institute and/or institutional collaborations, to explore and testify our ideas. The projects are categorised into Augmented Fabrication Research projects and Augmentation Design Research projects. While the former will focus more on a narrower concept of augmentation which deals with specific fabrication approaches using AR devices, the latter discusses more of the wider concept of augmentation, such as the relationship between the city environment and personalised human experiences.

Over the past decade, we have witnessed rapid advancements on both practical and theoretical levels in regards to automated construction and, as a consequence, increasing sophistication of digital fabrication technologies such as robotics fabrication and assembly, and large-scale robotic 3D printing. Most of this research, as it largely focuses on mass customisation, oversimplifies making processes, forcefully tweaking the material to follow unnatural assemblage, for which 3D printing is an example. There is a lack of exploration in natural material behaviours, which many delicate and complex crafting processes require. On the other hand, there exists a question of precision and efficiency when working with crafting techniques on an architectural scale. With this in mind, we focused our attention on the exploration of the ways in which traditional crafting techniques can be enhanced through AR technologies and given access to data previously exclusive to machines.

Case project 1: Steampunk Pavilion

The Steampunk Pavilion is constructed from steam-bent hardwood using primitive hand tools augmented with the precision of intelligent holographic guides. Designed by Gwyllim Jahn, Cameron Newnham (Fologram), Soomeen Hahm Design and Igor Pantic with Format Engineers, the installation was built for the 5th Tallinn Architecture Biennale 2019 in Estonia.

The pavilion explores how the application of MR environments could generate a new construction paradigm. The structure is a prototype for an adaptive design and fabrication system, resilient to wide tolerances in material behaviour and fabrication accuracy while being the largest structure to date built on the principles of AR-assisted fabrication. It explores alternative strategies for the fabrication of digitally designed architectural structures, utilising cutting-edge head-mounted devices (HMDs) to holographically assist workers in the manufacturing and assembly of highly varied components using traditional craft techniques. This enables the design process to be dynamic and the fabrication process to be adaptive, as the forms of parts can be changed at any time during the fabrication process without requiring any new tools, materials, drawings or code.

The timber elements in the structure are fabricated following the traditional process of steam bending. Each strip is bagged, steamed and bent over an adaptable formwork, using a holographic model as a reference to the desired result. The pavilion works with the constraint of two fundamental elements of 100 x 10 mm ash profiles and 30mm steel section, formed with the help of holographic models. This removes the necessity of anticipating every aspect of material behaviour in digital models, leaving open a certain degree of indeterminacy as material effects are discovered, desired and amplified during construction. It is this liberation of digital expression from the constraints of digital fabrication, together with the opportunity for nuance and material effects derived from material craft, that drives the architectural effects of the pavilion.

Apart from the novelty, the pavilion contributes to the architectural design discourse on multiple fronts: the discussion between digital and analogue, augmentation, technical innovation, as well as overlap of automation and handcraft. AR-assisted processes have the capability of enhancing human labour with data previously exclusive to machines, while enabling seamless inclusion of intuitive decision-making and experience, often absent from automated construction processes.

Above: Steampunk Pavilion, Tallinn, Estonia, 2019. The pavilion is built using AR technique alone. By Gwyllim Jahn, Cameron Newnham, Soomeen Hahm Design and Igor Pantic.

Below: Steampunk Pavilion, Tallinn, Estonia, 2019. The makers of the Steampunk Pavilion are bending the timbers based on the guideline from the hologram seen through AR lenses. By Gwyllim Jahn, Cameron Newnham, Soomeen Hahm Design and Igor Pantic.

Case project 2: Augmented Grounds

Augmented Grounds is a winning competition entry for the International Garden Festival 2020. It is a landscape design installation, located in the Jardins de Métis, Grand-Métis, Quebec, Canada. Construction was completed in July 2020. The installation celebrates human craftsmanship through a partnership with technology. The design is inspired by a traditional Métis sash, which is made with the art of finger weaving, and draped across one's shoulder or tied around one's waist. The Augmented Grounds garden represents the sash through colourful ropes made of twisted fibres that are tightly laid on top of the terrain to create a landscape of contours that reflects the different depths of Métis's history represented on the sash. While the experience of the installation is highly analogue, the construction process of this topographic terrain contributes to a new innovative practice of garden design by introducing smart construction technology using AR. As the geometry is generated, based on a mathematical algorithm, the combination of traditional materiality and mathematical form surrounded by the forest of the Jardin de Métis provides a unique experience for visitors to truly experience the product of collaboration between human, computer and nature.

The project utilised AR and cloud-based digital twin communication platforms to realise the construction during the global pandemic. It celebrates the fast and intuitive communication between designer and maker, utilising the digitally augmented workers who crafted the delicate material on site by wearing AR lenses. At the same time, a globally assembled team of designers was able to review the construction process through a cloud-based digital twin on the construction site, allowing them to intuitively supervise the construction process from a distance and efficiently pass on their knowledge and guidance to the local crew.

These two projects utilised cutting-edge technologies, such as AR devices and digital twin technology, to resolve the construction and communication issues in built projects. As the projects evolved, several unresolved technical issues (due to the limitation of current devices), as well as further potential for future application of AR technology, became apparent.

Real-time feedback, sensing, scanning, machine vision and cloud communications are some of the topics which were further explored in the academic environment through research projects presented in the following section.

Left: Augmented Grounds, Jardin de Métis, Grand-Métis, Quebec, Canada, 2020. Scene of a workstation by one of the designers supervising the local construction team remotely by reviewing the site progress via digital twin. By Soomeen Hahm, Yumi Lee and JaeHeon Jung.

Below: Augmented Grounds, Jardin de Métis, Grand-Métis, Quebec, Canada, 2020. The installation was completely built from remote training, thanks to the AR devices and online digital twin cloud communication platform. By Soomeen Hahm, Yumi Lee and JaeHeon Jung.

Augmented fabrication research – Beyond practice

There are still some remaining questions and yet-to-develop technologies central to this research. Within Bartlett's Research Cluster 9, starting from the 2016/17 academic year, we began developing a series of research projects that explore further potentials of augmented fabrication processes, specifically looking at various ways of dealing with human interaction and computer vision/sensor recognition.

Based on the technological difference, the AR Fabrication Research can be divided into three main categories:

1. *Human-data interaction:* ways in which humans interact with digital models based on the device gesture or eye-tracking sensors. This is fundamentally similar to any gesture interaction in VR.
2. *Human reacting to digital data:* ways in which humans utilise the digital data. In our case, it is used as a guide for construction.
3. *Computer reaction to physical environment:* ways in which computers recognise the ever-changing physical environment.

A number of tech companies have been competitively launching and exploring more realistic and intuitive human-digital interactions. For example, Microsoft has launched its second generation of AR device, HoloLens 2, with the iconic keywords of 'instinctual interaction' to describe its improved user experience. Compared to its earlier generation device (HoloLens 1), it provides a noticeably better tangible and intuitive experience for users, thanks to the hand- and eye-tracking system. If a project like Steampunk Pavilion, which utilised HoloLens 1, achieved a milestone in MR-assisted fabrication without the enhanced gesture- and eye-tracking systems, what will this latest technological advancement mean to architects, designers and builders? Does this improvement in

Future work will take into consideration the additional development on the aspect of real-time interaction between the physical and the digital.

technology have the potential to open new avenues in the way we design and shape our physical realities?

In our project Flowmorph, part of the design process included hand gesture input by a designer, paired with fluid simulation. While the designer 'draws' in physical space, the gestures are recorded and translated into a digital model which affects a fluid simulation. The resulting geometries can be used as a base data to form desired design products. In this case, the resulting vector field is translated into a simplified geometry set which becomes a guideline for the physical model.

A fundamental aspect of the current ongoing augmented fabrication research is the human reaction to digital data in the form of following holographic templates during the production process. This is important, as it utilises human intuition in its maximum capacity to deal with any unexpected environmental or physical malfunctions and issues, which makes the resulting product high in tolerance bearing.

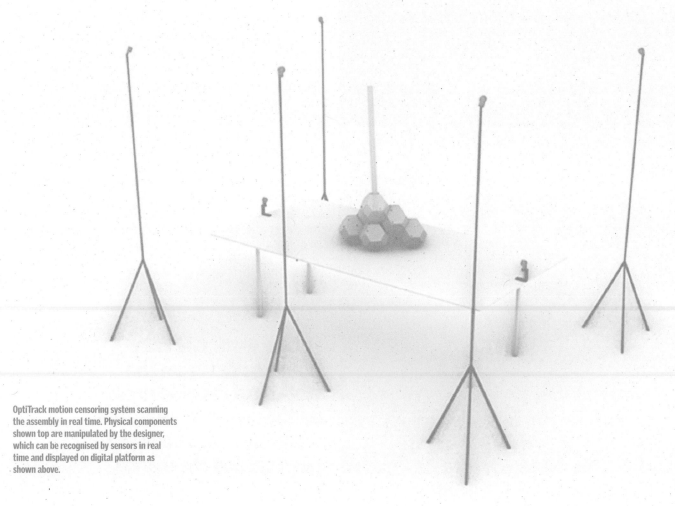

OptiTrack motion censoring system scanning the assembly in real time. Physical components shown top are manipulated by the designer, which can be recognised by sensors in real time and displayed on digital platform as shown above.

Computer reaction to physical environment: Reacting to real-time scanning

The nature of augmented fabrication is already allowing humans to make intuitive decisions during the making process and to make real-time changes to the design. Thus, it is important for a computer to recognise these human-made changes (or changes made by structural or environmental reasons) and to respond accordingly. However, although the current state of AR technology and devices does allow for real-time environmental scanning, these are still typically rudimentary models, lacking in detail and precision, and are not yet able to reproduce digital replicas of intricate models to a satisfactory level. In order to overcome this limitation and to create a streamlined feedback process between the real and the virtual, inclusion of real-time motion-tracking cameras such as OptiTrack or Kinect and techniques such as colour tracking and detecting and 3D depth scanning were needed.

The project BrickChain explored an assembly of simple 3D units, with specifically designed joints, allowing for the whole system to be assembled and disassembled during or after the construction process. Any changes in the physical structure would need to be fed back to the computer, in order for it to provide the next suggested guide as fast as possible. This means that the computer needs to recognise the physically assembled parts, which can be done through a number of ways, such as embedding custom sensors or Quick Response (QR) codes on each of the components. The project finally utilised a motion-sensing system by OptiTrack, which uses reflective material finish as sensors and reads the points from multiple camera positions and eventually calculates the 3D coordinates of each and every visible sensor point. The reflective sensors on the materials can then become part of the design, potentially aesthetically integrated into the surface finishes for better computer vision recognition.

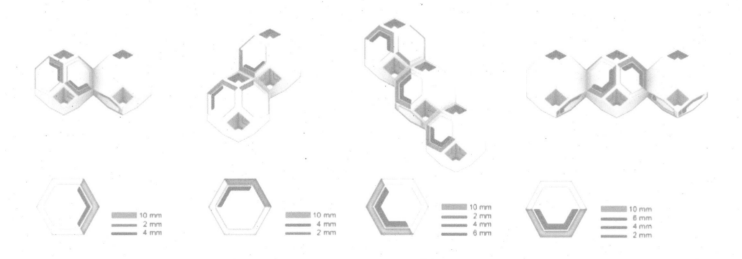

Reflective sensory materials are important elements to allow motion sensors to recognise the physical components, but they are also highly visible to human eye. This diagram illustrates the patterns designed for a dedicated reflective sensory material chamber which can also perform as a label for human builders.

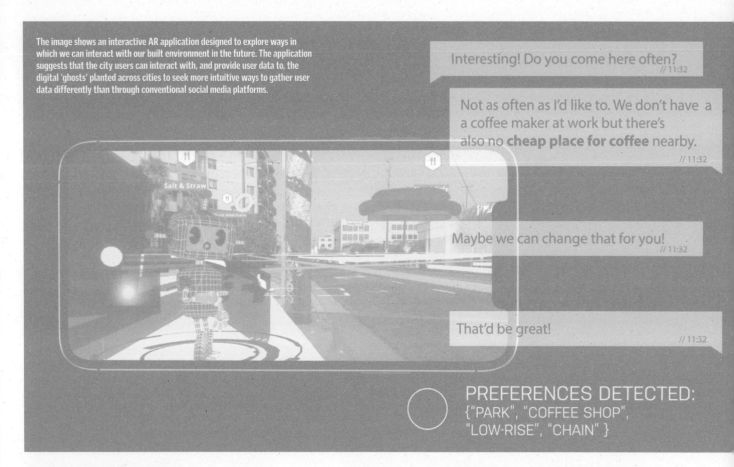

The image shows an interactive AR application designed to explore ways in which we can interact with our built environment in the future. The application suggests that the city users can interact with, and provide user data to, the digital 'ghosts' planted across cities to seek more intuitive ways to gather user data differently than through conventional social media platforms.

Interesting! Do you come here often?
// 11:32

Not as often as I'd like to. We don't have a a coffee maker at work but there's also no **cheap place for coffee** nearby.
// 11:32

Maybe we can change that for you!
// 11:32

That'd be great!
// 11:32

PREFERENCES DETECTED:
{"PARK", "COFFEE SHOP", "LOW-RISE", "CHAIN" }

Augmented design research –
Further research spectrum in augmentation

To further broaden the topic, we need to think about the impact of augmentation in a wider context and start exploring the relationship with cities and their social, economic and political issues.

Project Bellhop is a conceptual urban interaction social media project developed by Ian Dayagbil and Calvin Sin at SCI-Arc under the tutorship of Soomeen Hahm. Bellhop taps into the various streams of data that make up our geospatial world. Information such as land use type, land value, proximities and other metrics are brought together into one package that generates the Bellhop. Bellhops can be generated and bound to every known location, creating a network of digital beings across our cities. These digital beings gather data through 'normal' conversation and derive conclusions that can be used for refining generative design solutions.

Using this conceptual research project as an example, we have started to look explicitly at urban public spaces by simulating their adaptive and dynamic use. We have been seeking a potential smart system to dynamically occupy our urban environment, for which we developed an AR application that allows users to experience and interact with multiple agencies, which exist in both the real and virtual worlds, at the same time.

This aspect of research specifically focuses on how digital augmentation of the built environment is transforming how we inhabit cities and how they increasingly function as platforms. Smart cities, or cities which utilise digital platforms to manage utilities, physical plants and occupant use are a continually evolving concept. These platforms are evolving the way in which we inhabit the urban environment and they require us to reconsider their architectural implications.

The few studies showcased in this paper present the beginning of our research, which tries to bridge the gap between automated construction and human craftsmanship by introducing a workflow and tools which help transform humans into augmented human designers and builders. This trajectory will continue its development through a combination of AR technology as well as other technologies, such as wearable robots, AI and computer recognition, in order to advance the

modes of human-machine interaction while maintaining an intuitive and creative component to an increasingly automated chain of production. The aim is to develop an adaptable workflow and toolset, applicable to various design-to-fabrication scenarios.

Future work will take into consideration the additional development on the aspect of real-time interaction between the physical and the digital. Our final goal is to create an inventory of the culture of augmented making, identifying and dimensioning the effect of the many emerging technologies in the critical production of design and in the collective engagement of the designer in the reflective practice of making.

Conclusion

The above projects give an overview of the current state of MR technologies and the potential for their application in the field of architectural design and fabrication, through the lens of augmented craftsmanship and real-time participatory design. By doing so, they outline an approach which integrates digital augmentation into the cognition models of

making, challenging the traditional role of the artisan while essentially democratising the art of making.

The projects also enhance the relationship between technology, design and making by proposing a circular workflow instead of the traditional linear process. From the research approach, making is a tool for removing the barriers between digital design, digital fabrication and hand crafting. By utilising 'unexpectedness' in conjunction to materiality, it might be possible to design the unconventional or the unseen. The proposed new approach used in the introduced models considers factors and activities that are related to the external environment of designing (design medium) and to the different types and levels of representation.

1   King, B., *Augmented: Life in the Smart Lane*, Marshall Cavendish International (Asia), Singapore, 2016.
2   Kaku, M., *The Future of the Mind: The Scientific Quest to Understand, Enhance and Empower the Mind*, Penguin, London, 2015.
3   Coppens, A., *Merging real and virtual worlds: An analysis of the state of the art and practical evaluation of Microsoft Hololens*, master's thesis, University of Mons, 2017.
4   Hahm, S., Maciel, A., Sumitomo, E. and Rodriguez, A., 'FlowMorph: Exploring the human-material interaction in digitally augmented craftsmanship', Proceedings of CAADRIA 2019.

Given the urgency of climate change, the undoubted prominent challenge for architects today is to create a more sustainable and resilient built environment.

**Austrian Institute
of Technology (AIT)
City Intelligence Lab
(CIL)**

The advent of the computational design paradigm has brought a wide range of 'performance-driven' design methodologies to the forefront of architectural research and practice. These methodologies have brought along performance feedback mechanisms that were unthinkable a couple of decades ago. Today a computational designer can run complex – and often incomprehensible – analyses and simulations with a few simple clicks, such as finite element analyses for structural performance assessment, solar radiation analysis to evaluate exposure to the sun or even the notoriously complex computational fluid dynamics (CFD) simulations to assess the wind's performance in and around buildings. The unprecedented design intelligence that is embedded in these computational methodologies allows architects today to make much more informed design decisions.[1]

Given the urgency of climate change, the undoubted prominent challenge for architects today is to create a more sustainable and resilient built environment. Although these novel design methodologies did not necessarily come about to serve this purpose, the high level of performance control that they enable can evidently be used to respond to this challenge. Design decisions that significantly affect the environmental performance of a building, especially in the urban scale, are mostly taken in early stages of design. These early stages almost always require fast design cycles and do not lend themselves to time-consuming computational processes, such as CFD simulations that typically take hours to run. These complex simulations often require equally complex post-processing of their results and adequate domain knowledge to draw meaningful design decisions from them. Despite the continuous integration of building performance simulation (BPS) tools in computational design methodologies, the lack of fast, well-integrated and easily accessible environmental simulations is still the main barrier for integrating them in early stages of design, where they are crucially needed.[2]

InFraReD is an intelligent framework for resilient design, developed by the City Intelligence Lab (CIL) of the Austrian Institute of Technology (AIT), which aims to address the challenge by employing state-of-the-art artificial intelligence methods and large simulation datasets to enable fast, almost real-time environmental performance feedback in an open and accessible way. By integrating machine-learning models in mainstream computational frameworks, InFraReD makes environmental simulations accessible to designers, enabling the democratisation of environmental performance for buildings. Our initial findings on predicting key environmental performance metrics have been promising as a first step towards the development of an intelligent design-decision support framework that can promote resilient design methodologies at early stages.

InFraReD

InFraReD uses machine learning (ML) and specifically deep learning (DL) models that have been trained with large datasets of typically useful environmental performance simulations, namely solar radiation (SR), sunlight hours (SH) and computational fluid dynamics (CFD).[3] These models are trained to predict the simulation results in near real time and therefore minimise the typically large computational demand of performing these analyses. For example, instead of taking a few hours to run a CFD analysis to assess the pedestrian wind comfort of a single design iteration in a project, many iterations can be assessed interactively with immediate feedback on the wind performance. The ability to immediately understand the impact of a design decision with relation to the wind performance is unprecedented.

InFraReD's prediction capabilities were first demonstrated using SR simulation predictions. A typical SR simulation would take on average several minutes to run, while a SR prediction from one of InFraReD's ML models is produced in milliseconds. To achieve this, a large SR simulation dataset was used, which was produced with custom modelling, simulation workflows and the use of CIL's computing infrastructure. Thousands of SR simulations are run, and

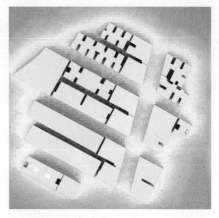

InFraReD's Machine Learning model predicting Solar Radiation (SR) simulations for a city block. Comparison of a normal SR simulation (above) with a ML prediction (opposite).

An SR simulation dataset used for training InFraReD's ML models. Thousands of typical SR simulations are used to train the ML model to predict the SR in milliseconds.

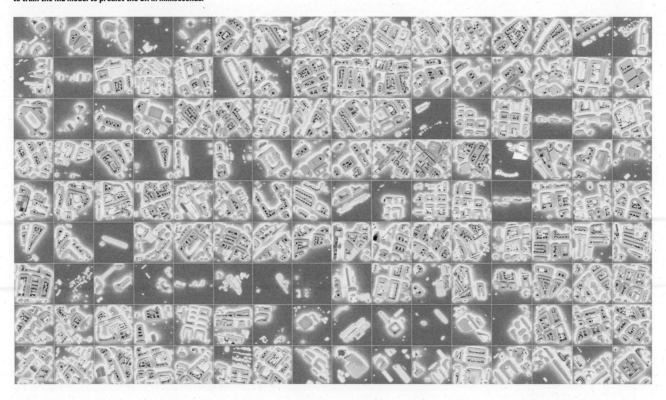

then their results are used to train the DL models to predict the effect of the building masses on the solar exposure. While SR simulations may not be as time-consuming and could be, up to a degree, run iteratively, CFD simulations are notoriously time-consuming and thus only used at late design stages for validation of design intentions. Within InFraReD, a CFD simulation that would normally take more than five hours can be run in a few seconds. An annual wind comfort study involves running multiple CFD simulations and calculating the occurrence of certain wind speeds for the whole year. A yearly comfort analysis that would take days to produce can be run in InFraReD within a few minutes.

It is evident that ML models can have a great impact in the integration of environmental simulations in the early stage design process. The accuracy of these models is of course not the same as an actual simulation, but in these early stages, when fundamental design decisions are taken, a small error – ranging between 5% and 15% – is negligible in relation to the potential gains in the achieved building performance. This technology is of course not new. Similar DL models are used on mobile phones and cameras to transform still or moving images; nevertheless, their application in empowering environmental simulations is just now being discovered.

Yearly wind comfort study using InFraReD. A yearly comfort analysis that would take days to produce can be run in InFraReD within a few minutes.

Embedding Performance Intelligence

A real-time environmental simulation model is a significant step in embedding performance intelligence in the design process, but its adaptation is questionable if it is not integrated in established design systems. InFraReD, as its name suggests, is not envisioned as a software or a tool but as a design and simulation framework that can be openly integrated and adapted to existing design systems and other computational frameworks. For that reason, InFraReD consists of both a back-end application programming interface (API) that hosts the ML simulation models and design data, as well as different front-end interfaces that allow users to interact with the ML models in a seamless and accessible way. These front-end interfaces include computational design tools (Rhino/Grasshopper) and custom scripts (Python), as well as open web interfaces and web-based design systems. InFraReD is also available on the CIL website[4] as an open access tool.

Further to these design interfaces, InFraReD is integrated in different interactive design systems, including augmented reality (AR) physical models, portable devices and interactive screens and tables which are all physically installed in the space of the City Intelligence Lab. The development of this ecosystem of interactive design interfaces is driven by the incentive to embed these highly complex performance feedback ML models in a natural and accessible way into the design toolbox of architects and planners. A physical model or an interactive screen is a much more natural way to discuss the diverse and often conflicting performance objectives of a project than a computer screen, and the immersive interactive environment of the lab has proven to be an engaging platform for discussion among the different stakeholders of urban projects.

The development of this ecosystem of interactive design interfaces is driven by the incentive to embed these highly complex performance feedback ML models in a natural and accessible way into the design toolbox of architects and planners.

InFraReD integrated within Rhino/Grasshopper. It provides ML prediction of CFD within the most commonly used computational design framework.

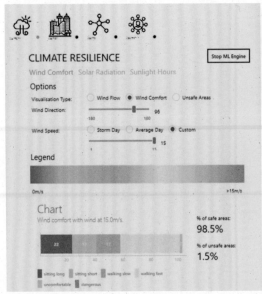

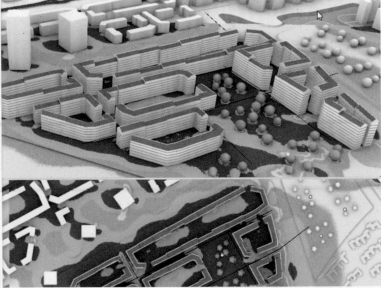

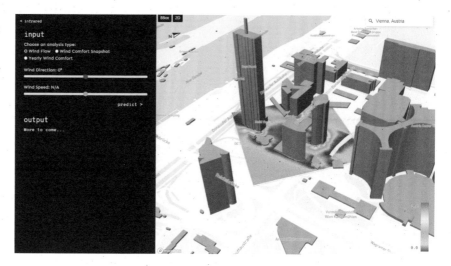

A CFD prediction on the InFraReD website. InFraReD is an openly accessible platform for real-time environmental performance.

InFraReD's different interactive design interfaces – an open-ended system for performance-driven design.

An intelligent performance-driven design system needs to inevitably integrate all aspects of design performance to provide a holistic understanding of the intertwined parameters and objectives of an urban project.

Different performance analyses integrated in InFraReD.

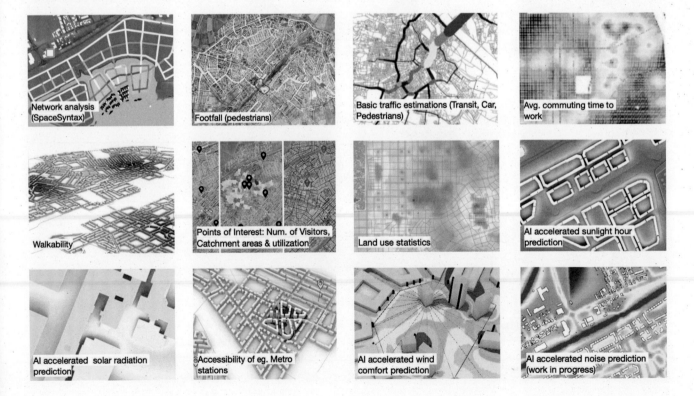

Network analysis (SpaceSyntax)

Footfall (pedestrians)

Basic traffic estimations (Transit, Car, Pedestrians)

Avg. commuting time to work

Walkability

Points of Interest: Num. of Visitors, Catchment areas & utilization

Land use statistics

AI accelerated sunlight hour prediction

AI accelerated solar radiation prediction

Accessibility of eg. Metro stations

AI accelerated wind comfort prediction

AI accelerated noise prediction (work in progress)

An augmented physical model using InFraReD at the CIL creates an immersive and informative design decision platform.

Exploring Design Spaces

InFraReD is not focused only on environmental performance. An intelligent performance-driven design system needs to inevitably integrate all aspects of design performance to provide a holistic understanding of the intertwined parameters and objectives of an urban project. Further to the environmental metrics, key performance analyses – such as spatial metrics, noise propagation, accessibility to transport and infrastructure or pedestrian dynamics – are all integrated in InFraReD's ecosystem, allowing its users to explore vast solution spaces of diverse objectives. It is envisioned as an open system that aims to expand in all aspects of urban performance to allow its users to intelligently negotiate the different performance goals and design parameters, and to reach unprecedented urban solutions.

One common misconception that the computational design paradigm has brought to the architectural discourse is that design control is slipping from the hands of the architect and is being handed over to the black boxes of the performance-driven algorithms. Whether seen from the scope of the architect, an environmental specialist or a city planner, these black boxes almost always seem to take away the ingenuity and the empirical knowledge of the human. The theoretical and research framework of InFraReD aims to do exactly the opposite. It aims to empower the architect, the planner and the specialist to have a more informed dialogue and collectively make better decisions by augmenting their ingenuity with artificially embedded intelligence.

1   Clarke, J.A. and Hensen, J.L.M., 'Integrated building performance simulation: Progress, prospects and requirements', *Building and Environment*, vol. 91, 2015, pp. 294–306.
2   Purup, P.B. and Petersen, S., 'Research framework for development of building performance simulation tools for early design stages', *Automation in Construction*, vol. 109, 2020, pp. 1–15.
3   Nahan, R.B., *Architect's Guide to Building Performance*, The American Institute of Architects, Washington D.C., 2019.
4   http://infrared.city/

PROFILE:
Disrupting Design(ers) through Automation
Bryden Wood

We are not software developers, we are architects and engineers, but we do make software. When someone wants to join the Creative Tech team at Bryden Wood, we always ask them the same question: 'Do you want to design with digital technology, or do you want to make digital technology for others to design with?'

There is no right answer to this question – and some people have changed their answer over time – but it helps us to understand the starting point for a new member of the team. It also helps to immediately situate our own starting point, which is that we are using our knowledge and training as designers of buildings and infrastructures in order to build tools for other people. We want to allow as many people as possible to be creative and participate in the design of the places in which they live, work and move through.

We are using 'design automation' to make digital tools and technology that provide 'intelligent control' in design and allow people to be creative and practical when proposing social or economic infrastructure.

The software we use in our industry is, somewhat understandably, not targeting a specific functionality. Instead, it is trying to appeal to a wider audience and to deliver functionality to the broadest cross-section of users. This means that it will often support commonly understood or mainstream design behaviours but will not push the boundaries of what is possible. It is not the place to find the future of design; it stabilises recent or novel design approaches, but it does not help develop new ones.

We are not trying to re-build CAD or BIM tools and nor are we trying to re-imagine those paradigms. By using the word 'automation' we do not just mean faster, we mean smarter. And by using the word 'smarter', we do not just mean algorithms for the sake of algorithms. We want to use design automation to create outcomes that are fundamentally different when compared to other design processes.

We have developed methodological and technical systems for design automation across a range of different technologies that embed the principles of Design for Manufacture and Assembly (DfMA) and mass customisation to radically change the nature of design and to challenge our own roles as designers within it.

We are continually weaving together four threads that connect our work. They are not linear, and they are not the steps in a process; they are continuous and reflexive. The relationships are not fixed but are knitted together in different ways across time so that they are strengthened by each other and are able to hold themselves and resist tension:

1. *Patterns* (our synthetic reason): we look for the patterns that describe how design systems and processes work and we turn them into machine-readable logics and rules.

Platforms. Pre-engineered physical components can be handled as data points – rather than geometry – to auto-generate lightweight models that are the instructions for the digital (and physical) assembly processes that follow.

2. *Algorithms* (our intelligent processes): we use different algorithms to suit different patterns, but we are cautious. Algorithmic processes can be complex and not easily understood, so we look for moments of intervention within these automation process that allow people to interfere and hopefully understand the process (and we document everything!).
3. *Connections* (our distributed technologies): we do not exist in a bubble, so we join our tech to other systems and platforms so as not to duplicate functionality where it already exists and to avoid centralising either data or process.
4. *Interactions* (our digital companions): ultimately, technology is about people and we want to put design automation into the hand of the many. The experience is not supposed to be deterministic but reflexive and co-created between people and technology.

Through design automation we not only want to change the way people think about architects and engineers but also how we think about ourselves. The topic of automation is somewhat controversial and is often seen as a threat – both to processes and outcomes. While there are certainly instances and examples of negative consequences of the application of automation technologies in other industries, the risk is surely greater if the push towards automation is brought from outside. The danger of a disruption – of a shock to the system – is only increased if the pressure is external. We are trying to encourage change from within and to encourage architects and engineers to be creative with digital technology, not only to consume it or, worse, to become victims of it. In this way we can collectively re-imagine what it means to be a designer and to design tools for others as well as ourselves.

We are not software developers, but we do make software. In the projects that we have worked on – a short representative selection of which are described below – we have had to learn a lot of new technologies, skills and behaviours. And we have made a lot of mistakes. Making software is not necessarily easy (at least not for us) and it is often messy, but we are not aiming for the normative attitudes of software development and nor are we aiming for those outcomes. We want to create technologies for others and to share them so that, through design automation, we can realise a better built environment.

We want to create technologies for others and to share them so that, through design automation, we can realise a better built environment.

Many different patterns can be created for any given system and so they are temporary – not fixed – representations of our understanding and interpretations of the underlying systems of design.

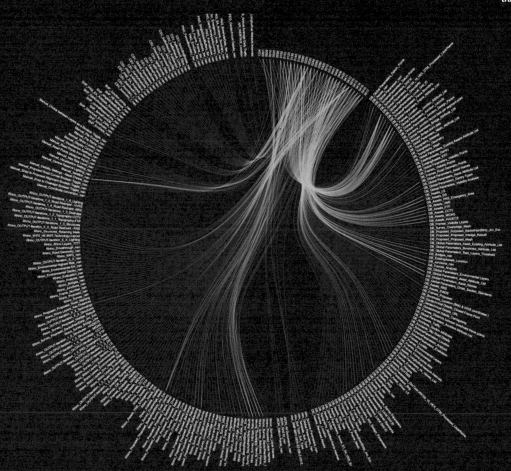

Connections. The ways in which things are connected affects the behaviour of the system as a whole. There is no single solution to this, but rather it is a process to find and make the right connections. The goal is not (necessarily) stability but, instead, transparency.

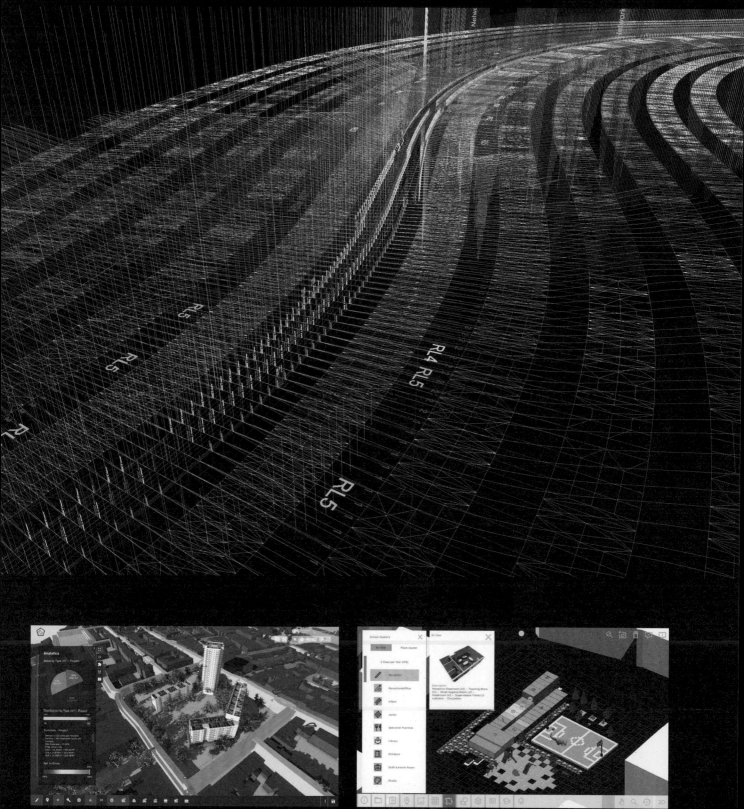

Automating Design for Manufacture and Assembly
Configuring buildings

PRiSM app was developed by Bryden Wood in collaboration with
Cast Consultancy and the Greater London Authority (GLA), as
part of the Mayor's initiative to increase the adoption of Precision
Manufactured Housing (PMH) across London. This open source web
app was developed to share manufacturing know-how with designers
(architects/developers/housing associations/councils), to increase
awareness regarding the opportunities for precision manufactured
housing techniques. Users can rapidly model their apartment scheme
within a 3D context of London, in accordance with their specific
development brief (key apartment dimensions and apartment mix).
The app then provides visual feedback regarding available PMH options
(such as volumetric and panelised systems) which are suitable for the
specific design, and how to amend the design to make the remaining
systems viable. The design functionality within the app was informed by
a study of open source industry datasets relating to housing schemes
built within London, as well as project data submitted by project
stakeholders, and this research was published to accompany the app.

SEIMSIC is an open source and free to use web app which allows users to
configure spatially compliant school building blocks (known as 'clusters')
within a 3D site context anywhere in the UK and design a primary school
in minutes. The clusters were developed in collaboration with the
Department for Education (DfE) to meet its schedule of accommodation
requirements, meaning that designers do not have to re-interpret design
guidance every time. The app seeks to democratise the school design
process with its game-like interface, meaning that anyone (teachers/
parents + children) can be involved in the school design process.

Generating infrastructures

Rapid Engineering Model (REM) is a digital toolkit developed with
Highways England to bring automated design to the strategic road
network. REM combines topographic and environmental analytics
with encoded design rules to auto-generate many kilometres of road
enhancement in a fraction of the time required for manual design
processes. Different design scenarios can quickly and easily be
explored by the users to test optimisation strategies and to understand
the impact on sustainability, constructability and safety. REM even
allows the rules themselves to be tested, to demonstrate the effect
of different constraints on design outcomes. Using the REM toolkit,
consistent and transparent design outcomes – including both data
and geometry – can be easily generated and shared to allow the design
teams to make better decisions with more transparency.

Railway: Automated Infrastructure Design (RAID)

RAID is an open-source app which accelerates the design process for track alignment within the UK railway network. The tool uses a generative design technique combining environmental data and engineering rules. It provides the ability to quickly assess the rail network and explore the performance of different route options at a macro scale. Macro level solutions can be generated for completely new routes and track alignments or explored along existing railway corridors. Users create multiple scenarios which can be explored through a series of interactive data analysis tools.

From digital to physical

PRiSMs to Platforms: Our Platform Automation workflow takes early stage design configurations that have been created in PRiSM and generates a sophisticated design model using detailed parts, components and assemblies from Bryden Wood's own 'Platform' construction system. The kit-of-parts contains architectural, structural and building services elements, and results in a data-rich, pre-engineered and coordinated DfMA BIM model. This workflow employs procedural design automation processes to leverage commercially available BIM tools to create content that can be used to accelerate the delivery of residential projects.

Framework for Robotics and Automated Construction (FRAC) is the automation and robotics strand of our Platform Approach to DfMA. FRAC is a digital toolkit that connects commonly used design software to robotic technology tools, using a suite of custom-built apps. We are using FRAC to design building components optimised for robotic manufacture from the outset, rather than trying to re-engineer manufacturing processes to suit existing parts. FRAC provides an open, extendable and collaborative platform that enables all designers to engage with robotic technologies and to embed digital manufacture workflows into their everyday design activities.

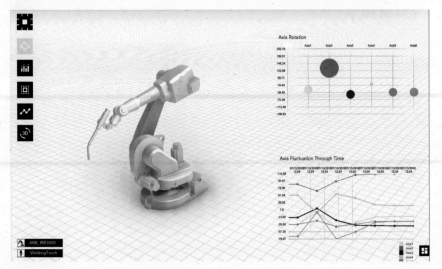

FRAC. We build interactive and immersive tools to allow designers to 'encounter' robotic technology as part of their everyday practice and to help them embed it as part of the natural, iterative process of design.

Through design automation we not only want to change the way people think about architects and engineers but also how we think about ourselves.

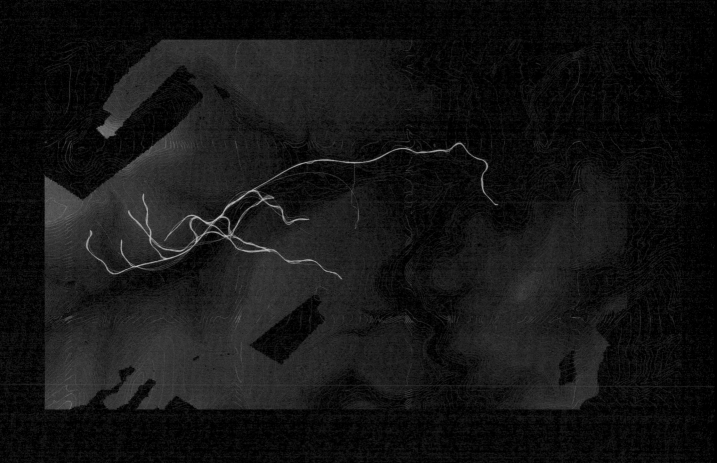

Engineering rules and geo-spatial data analytics are combined to quickly generate many possible track alignments from which the most suitable can be selected.

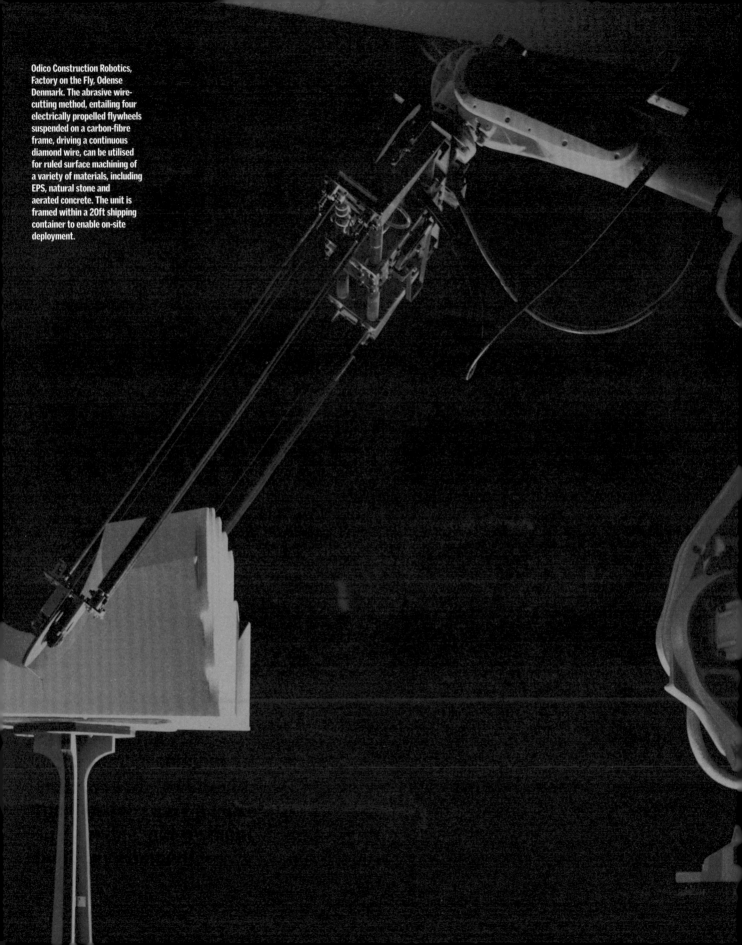

Odico Construction Robotics, Factory on the Fly, Odense Denmark. The abrasive wire-cutting method, entailing four electrically propelled flywheels suspended on a carbon-fibre frame, driving a continuous diamond wire, can be utilised for ruled surface machining of a variety of materials, including EPS, natural stone and aerated concrete. The unit is framed within a 20ft shipping container to enable on-site deployment.

PROFILE:
Scaling Construction Robotics
Odico Construction Robotics

Odico Construction Robotics was founded on the belief that the broader imperative to enhance the built environment's sustainability holds a positive driver towards enriching the quality of architectural design. One key component of this notion is the widening theoretical body documenting an inverse proportionality between formal complexity and the consumption of construction material. This relationship can be explored through computational techniques and forsees a design future remarkably different from common extrapolations of late modernism. The materialisation of this potential, however, is prohibited by the inflexibility of manual construction. The initiating moment for the founding of Odico was the realisation that relaxing these constraints through robotic technology can enable the arrival of this anticipated reality.

This chapter outlines the principal journey of Odico by describing the first technological innovations and current developments, and concludes by projecting forward with new fabrication innovations.

The first technological developments were fuelled by the observation that the global concrete industry is the largest subsector of the international construction sector, where the singular largest expenditure of activity in this space is – for the case of non-standard structures – formwork cost. The impending cost structure stems from the time-intensive manufacturing processes, which falls largely into two groups:

1. Manually machined and assembled timber moulds.
2. CNC-milled formwork blocks.

Olafur Eliasson and Sebastian Behmann with Studio Olafur Eliasson, Fjordenhus, 2009–18, Vejle, Denmark. Commissioned by Kirk Kapital. The project represents the first example of a commercial construction project in which the load-bearing concrete structure is erected using robotically wire-cut EPS formwork. Photo: Anders Sune Berg, © 2009–18 Studio Olafur Eliasson.

In the former, the many carpentry hours drive the cost, while in the latter, the driver is the excessive machining times stemming from milling requirements at 0.2mm step-over values. Effectively, demanding a tooltip to traverse thousands of times for 1m^2 formwork to achieve industry-standard smoothness levels.

However, most architectural surfaces conform to single or doubly ruled surfaces. Exploiting this geometric property opens the possibility of swiping the surface once when using a digitally controlled abrasive or heated wire. In digital fabrication, the single most determining expenditure is machining time as a function of depreciation of the machining hardware. Hence, a dramatic reduction in processing time for 1m2 of custom formwork surface holds the potential to achieve a proportional decrease of formwork costs. In practice, this translates to a reduction from several hours of CNC machining to a few minutes of digital wire-cutting for the same formwork geometry.

Armed with this knowledge, Odico aimed to transform the concrete construction by decoupling design complexity from formwork expenditure through the commercialisation of robotic hot-wire-cutting of Expanded Polystyrene (EPS) formwork. This was implemented through a case-based, iterative working philosophy of launching commercial pilot production from academic lab prototypes, hereby growing a financial base for continuous technology research in process engineering and control software domains to bring the concept to industrial scale. As such, real-world production would become the primary driver of technology innovation in a perpetual prototyping process of increasing scale and complexity, while reversely functioning as a live test-bed for new methodological developments.

Real-world production would become the primary driver of technology innovation in a perpetual prototyping process of increasing scale and complexity, while reversely functioning as a live test-bed for new methodological developments.

Aarhus School of Architecture & Odico A/S, Opticut Prototype, Aarhus, Denmark, 2018. In exploration of the boundaries set by the manufacturing constraints of robotic wire-cutting, Odico explored with multiple partners the realisation of topology optimised concrete structure via EPS formwork.

Over the course of increasingly scaling production projects, Odico was delegated the commission in 2013 – one year after its initial launch – to produce 4500m2 of doubly ruled EPS formwork for the Fjordenhus in Vejle, Denmark, designed by Olafur Eliasson and Sebastian Behmann with Studio Olafur Eliasson. As the first known example of a commercial building project with its concrete load-bearing structure executed via robotically wire-cut formwork, the project enabled a full-scale validation across its four-storey elevation of the industrial capacity of this manufacturing methodology. Following this breakthrough, an accelerated commercialisation ensued, delivering over 300 projects in seven countries with leading design offices such as Zaha Hadid Architects, Skidmore, Owings & Merrill (SOM), Bjarke Ingels Group, 3XN Architects and Dorte Mandrup. Following the first commercial successes, it became clear that to accelerate technology development at the pace required for full generalisation, a transition was needed for a new mode of operation.

The Factory on the Fly

This is a general-purpose technology platform that has been materialising since 2018 and is designed for on-site and in-factory construction robotics. Now, two years later we see the first factories being sold to private companies and being used on site. The Factory on the Fly consists of a transportable, cloud-connected robotic manufacturing cell hosting a plurality of machining processes and operated from a simple tablet interface. By allowing for both off-site parametric generation of production files and on-site operator customisation of work items, the Factory on the Fly enables unprecedented manufacturing flexibility levels.

Skidmore, Owings & Merrill LLP, Stereoform Slab, West Loop, Chicago, 2019. The potential of using ruled surface rationalisation to conform shape-optimised designs for realisation with robotic wire-cutting was further explored in collaboration between SOM and Odico in the context of the Stereoform Slab pavilion, indicating a 20% reduction potential for the material consumption of high-rise slabs. © David Burke.

A case for robotic timber

While the founding history of Odico has been closely linked with wire-cutting of EPS formwork, the Factory on the Fly constitutes a general-purpose, process-agnostic, cyber-physical platform technology based on a modular architecture, which can facilitate a plurality of machining processes. An extension of this capacity, Odico is currently exploring complementary means of production, including timber.

The case for timber provides an opportunity to revisit conventions of the man–machine interface. Currently, the Factory on the Fly communicates construction information to and from the robot via tablet interfaces. However, in our digitally saturated world we should remember that in the thirteenth and fourteenth centuries construction information for jigs or building components were communicated through 1:1 drawings that were etched into the plaster floors of on-site drawing houses made for the architects/builders.[1] Today, in conventional timber construction, hand-drawn symbols are still used to communicate the positioning of elements within a timber structure.

This begs the question: what if construction information for digital fabrication systems could be communicated via a pen and a ruler? This question is of great importance. Since the workers in the construction sector constitute crafts-trained carpenters, masons and concrete workers with limited exposure to robotic fabrication technologies, we seek to envision machine interaction methods that support human intuition.

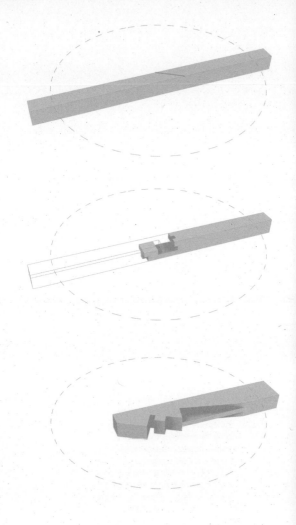

Top row left: Odico Construction Robotics, timber structure for a house, Odense, Denmark, 2020. A detail from the house, showing an example of applied parametric detailing that aids assembly by making it simpler to position and nail the structure together.

Top row right: Odico Construction Robotics, timber structure for a house, Odense, Denmark, 2020. A detail of position notches, which were robotically milled based on a parametric model, thereby minimising potential errors in on-site construction processes.

Bottom row left: Jens Pedersen, building inspection, Saksild, Denmark, 2019. The image documents that hand-drawn symbols are still used to communicate construction information today.

Bottom row right: Jens Pedersen, PhD documentation, Odense, Denmark, 2019. Imagine if you could programme a robot through something as intuitive as a pen and a ruler? The photo illustrates a not-so-distant future where one draws on workpieces instead of writing text-based instructions.

Left: Jens Pedersen, PhD diagram, Odense, Denmark, 2019. A diagram describing how the physical drawing, after being digitised, can be augmented with geometric or fabrication information, thereby allowing complex timber joints to be made from a simple starting point.

Right: Odico Construction Robotics, timber structure for a house, Odense, Denmark, 2020. Denmark's first commercial robotically fabricated timber house, a project that came to fruition through a close collaboration with BRAV Engineering and sought to develop a construction system that simplifies the assembly of conventional timber structures.

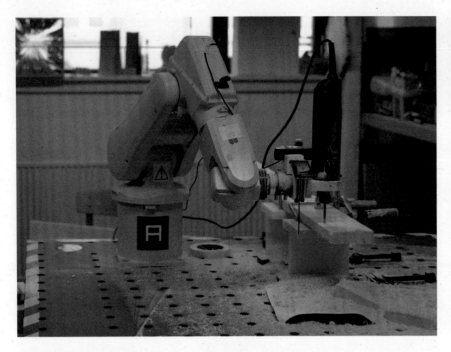

Drawing as an Interface

In the research project, Parawood, led by industrial PhD Fellow at Odico, Jens Pedersen explores a mode of operation where physical notations/markings made on workpieces are used to communicate construction information directly to the robot via a computational framework containing computer vision and machine learning algorithms. Marking(s) are detected by a camera mounted on the robotic arm and digitised by the computational framework. The markings can be symbols, lines, curves or numbers, as seen in contemporary carpentry praxis. The digitised drawing is visualised on a tablet for user validation and/or parametric augmentation. Augmenting the digital marking twin enables carpenters to generate geometrical/fabrication information necessary to make, for example, a French lock or other complex timber joints through a single line or curve drawn on the workpiece.

The method turns current digital fabrication paradigms upside down, since it shifts the starting condition from a digital model to a physical drawing. Additionally, the method has the benefit of removing a significant constraint of robotic fabrication: reach. In timber construction, workers often need to process timber at both ends, which in robotic terms would require either a large robot or an extra axis of freedom. This method allows the users to simply draw where they want the robot to process and then position the workpiece where the robot can 'see' the drawing. Therefore, the benefits of this method are twofold:

1. It can simplify the design of future Factory on The Fly systems.
2. It presents a latent potential for fields in education or craft professions with an interest in robotics.

The method turns current digital fabrication paradigms upside down, since it shifts the starting condition from a digital model to a physical drawing.

Far left: Jens Pedersen, PhD diagram, Odense, Denmark, 2019. The robot set-up, which was used to develop the 'drawing interface' method. It is a small ABB 120 robot, with a limited reach, which was used to create a wooden structure where pieces were far bigger than its reach.

Left: Jens Pedersen, PhD diagram, Odense, Denmark, 2019. The robot milling along a drawn line, where one can see the precision of the system.

Right: Jens Pedersen, PhD diagram, Odense, Denmark, 2019. A timber structure that was made using the 'drawing interface' method. The structure tested the precision of the system and the process of augmenting physical drawings with complex fabrication and joinery information. Every joint in the structure was made within a millimetre of the original drawing.

The technological barrier

Using robots requires a willingness to write lines of code, especially if you want to use large industrial robots. Through personal teaching experience in the field of architecture, this tends to have a discouraging effect on many architectural students, which arguably could be because coding is seen as a steep 'price of entry' into the realm of robotic fabrication. Therefore, how we teach robotic fabrication plays a key role for how technologies we develop at Odico can be implemented in architectural practice. We argue that the 'Drawing as Interface' method is a step in the right direction, as it lowers the technological barrier from coding to physical drawings and tablet use. Therefore, if the education of robotic fabrication becomes commonplace, we could foresee new types of practices, or design processes could emerge, as the prototypical phase of architectural design could be sped up and scaled up.

Extrapolating from these developments, we project a digital fabrication future, in which the facilitating technology becomes so aligned with human craft-intuition that a distinction between the two is virtually imperceptible. This merger, in turn, may help to overcome remaining perceptual barriers to tacit knowledge programming, enabling the emergence of a new cybernetic culture of trade-augmentation, which will catapult forward new generations of shapers and makers into a design-centric epoch of fabrication computing.

1   'The Tracing Floor of York Minster', in Studies in the History of Civil Engineering, 1:81–86, *The Engineering of Medieval Cathedrals*, Routledge, London, 1997.

The Living, bio-welded column at 1:1 scale, New York City, 2019. Test of full-scale bio-welded mycelium units.

PROFILE:
Natural Intelligence: Design with Living Materials
The Living

The intersection of biology and design reveals opportunities for a better built environment – one that is carbon neutral, locally sourced and adapts to change. Traditional design methods, however, are poorly suited to working with living materials, so designers must embrace new approaches that integrate complex systems, engage uncertain outcomes and leverage natural intelligence. At The Living, we explore how novel biological and computational approaches can integrate with a research-based practice and pedagogy. Our team comprises architects, educators, scientists and makers who share a relentless curiosity in exploring the edges of possibility and a willingness to take risks. Our design methods encompass both the ecosystem (the social, cultural and environmental context that surrounds and shapes) and the prototype (the materials, structures and forms that emerge in response).

Bio-manufacturing

Our interest in bio-manufacturing processes and biomaterials stems from the industry's growing negative environmental impact, the urgency of rapidly decarbonising our society and the need to seek alternatives to carbon-intensive materials. Natural factories – organisms and processes that have generated stable, structural materials for millions of years – may be a key component of decarbonised architecture. Mycelium, the root-like branching structure of fungi, is a prime example of how our thinking about materials can shift from extraction and refinement to cultivation and harvest. The natural behaviour of mycelium is to proliferate and knit together the loose, organic material of a forest floor, and it can be repurposed to convert any cellulose-based substrate (plus water and nutrients) into robust, structural, fire-resistant and 100% biodegradable blocks. To stop growth, the organism is rendered inert via heat-treatment (or desiccation).

The first grown tower

Hy-Fi, our winning entry to the 2014 MoMA Young Architects programme, was our first prototype of mycelium-based architecture. It demonstrated the properties and possibilities of this new biomaterial to create a building-scale installation designed to disappear without a trace. The 12-m tall tower combined the traditions and detailing of masonry with a material so unlike its heavy, compressive counterpart that it required Arup, our structural partner, to develop novel mycelium material models for their structure analysis software package GSA solver. Hy-Fi succeeded in not only exposing its audience to biomaterials but also in creating a tangible experience to see, feel and occupy a mycelium structure. The experiment also challenged our own assumptions that mycelium could replace conventional

At The Living, we explore how novel biological and computational approaches can integrate with a research-based practice and pedagogy.

masonry without changing the way we design or construct. Naturally hygroscopic, the bricks would expand and contract, an effect that when viewed in time-lapse made the tower appear to subtly dance in rhythm with the day–night cycle. These unexpected discoveries deepened our curiosity about the material's potential. Could we remove all inorganic elements, such as mortar? What would happen if the material itself remained active and alive?

Living Bricks

Hy-Fi's success led to a 2019 invitation to produce a new mycelium work for 'La Fabrique du vivant' at the Centre Pompidou, an exhibition exploring works that link living matter and technology. Our proposal, Living Bricks, was to create a captivating structure to greet visitors entering the exhibit that would showcase both the familiar (the form of a triumphal arch) with the unfamiliar (an experimental construction technique). Rather than stereotomy and carefully stacked masonry, our arch would self-assemble via two novel techniques: bio-welding and jamming.

Bio-welding

As the mycelium blocks grow, their fibrous networks gradually seal off their interior, leading the organism to densify along the exterior where oxygen is still accessible. We discovered that two or more living units, when placed in physical contact, would continue to grow and effectively graft themselves into a single continuous object. This process requires two stages:

1. An initial growth within a mould – to solidify the loose material.
2. A secondary growth out of the mould and aggregated in a larger formwork.

We repeatedly tested the process – varying growth time, unit shape and volume – and found the process reliably produced a solid, stable lattice of bio-welded units with no added material or bonding agents.

Jamming

Granular jamming is both a complex and everyday phenomenon. You experience it any time a sugar shaker, inverted over a cup of coffee, fails to emit any sugar. Individual particles of any granular material (from microscopic sugar crystals to macroscopic mycelium blocks) will typically behave like a fluid and flow with gravity when set in motion, but when subjected to compressive forces they seize up into a mass that behaves like a solid. Compression can be triggered by both self-weight and contact force with a container, so the transition can be *designed* by changing the properties of the particles and the geometry of the container. The phenomenon is well studied within academic physics and industry, and in the past decade some interesting

Top: The Living, Hy-Fi, Queens, New York City, 2014. A tri-lobed branching tower made of 10,000 grown and biodegradable bricks that was exhibited at MoMA PS1 and composted afterwards. Image courtesy of Amy Barkow.

Above: The Living, bio-welding experiment, New York City, 2019. Close-up of mycelium tissue grafting across adjacent mycelium units.

Right: The Living, jammed arch, New York City, 2019. 2D form-finding experiment via jamming of dried mycelium units.

Below: The Living, jamming experiments, New York City, 2019. Initial experiments to explore influence of formwork parameters on resulting jammed structures.

precedents have shown its application in construction, including the dome structures by Kentaro Tsubaki, wall structures by Karola Dierichs and Achim Menges, and column structures by Gramazio Kohler with MIT's Self Assembly Lab. Each precedent used the technique to create reversible structures, whereby releasing the compression would recover the units. Our prototype looked to do something unique, using the method to cast in place a solid, spanning structure made permanent through bio-welding.

Design of experiments

The sheer scale of the installation – 4m wide x 4m tall x 2m deep – precluded any full-scale test before installation. As an added challenge, we were developing the design and process in New York City but would have to fabricate in Paris. Facing something both ambitious and unproven, we designed a sequence of experiments to build confidence that the rules and process would produce the right outcome.

Small-scale experiments

We discovered the arch form while reviewing industrial research on jamming. Vessels such as funnels, hoppers and silos are susceptible to forming elegantly curved bridges of material over their outlets, and a significant amount of research has been devoted to mitigating this effect. We were interested in the opposite: deliberately triggering it and designing the form by tuning the system's parameters. Our first set of experiments used laser-cut wood particles and a reconfigurable acrylic vessel to find candidate formwork that would reliably produce the right arching structures.

Above: The Living, jamming experiments at 1:4 scale, New York City, 2019. Test to explore jamming and particle release procedures using mycelium units.

Medium- and large-scale experiments

The next experiments employed digital and physical methods to develop a particle shape that would successfully create jammed arches and bio-weld into a solid structure. Working with Professor Heinrich Jaeger of the University of Chicago, an expert in simulating granular jamming, we developed a shape optimisation to meet both fabrication and behavioural constraints. Our system discovered that slightly tapered cylinders, capable of being grown in ready-made vessels, performed nearly as well as the complex shapes used by precedent projects – and at a fraction of the cost. Ready-made moulds were also reusable after fabrication, eliminating a tangential waste stream of one-off moulds. Candidate designs were then physically tested at two scales:

1. A 1:4 simulation of the full assembly using thousands of units.
2. A 1:1 test of full-scale units in a bio-welded 1-m tall column.

The result was a detailed set of guidelines ranging from the optimal time and temperature for growth to the procedure for releasing the loose particles to reveal the jammed arch.

Full-scale experiment

We partnered with a design studio at ENSCI, a French national school of industrial design, to realise the carefully coordinated experiment. In a series of hands-on workshops, we taught the students about the biomaterial and trained them to grow the units and assist in the installation. As the mycelium was growing, the towering formwork was assembled and prepared for filling in the gallery. Over the course of one afternoon and evening, the 1,600 units were removed from their moulds and carefully poured into the form. In a tense moment, we delicately released the loose bricks to reveal the self-supporting arch. The structure was then left undisturbed for five days to bio-weld. Finally, the formwork was removed, revealing a towering arch of mycelium units suspended and spanning between two sloped pedestals.

Below: The Living, Living Bricks, Paris, Centre Pompidou, 2019. The final experiment at 1:1 scale is a jammed, asymmetrical and self-supporting arch of bio-welded grown units. Image courtesy of Andres Baron.

The future of design with living materials

Living Bricks was a new and tangible experiment in living architecture. Units were simultaneously brick and mortar, and the precision of stacked masonry was replaced by a self-assembling construction where the form emerges from stochastic probability. Our project tested and proved a novel form of construction, while also showcasing a biomaterial with almost no carbon emissions that was composted after the exhibition, leaving behind nearly zero unrecovered waste.

Faced with a global demand for a staggering quantity of new buildings and infrastructure needed to support a growing population, architects must avoid perpetuating the status quo of architecture

A future with living buildings is closer than we think.

as a major contributor to the shared crises of energy consumption, resource depletion and waste. In this context, living materials could play a critical role and offer viable alternatives to carbon-intensive materials and construction systems. Wide adoption will still require more research and rethinking the norms of building. We see this as an opportunity to foreground research as practice to discover new possibilities through experimentation. All buildings are inherently prototypes and opportunities to test, validate and learn, and to push the discourse and profession forward. By actively testing prototypes at scale and in the real world, we can demonstrate that a future with living buildings is closer than we think.

Below: The Living, Living Bricks, Paris, Centre Pompidou, 2019. Under-side of the arch, revealing the stochastic nature of the self-assembled mycelium units. Image courtesy of Andres Baron.

Acknowledgements
Living Bricks was created in collaboration with Columbia University GSAPP, ENSCI, Blast Studio, Centre Pompidou, Arup (Matt Clark and Shaina Saporta), Heinrich Jaeger, Cecil Barnes, Angelica Yannoulatou, Laelia Troubat, Clément Chapalain, Guillaume Rousseau, Gradinarov Alexandre, Benjamin Castera, Meganne Lelay, Samba-Camille N'Dao, Mathias Vives, Louise Badiane, Beth Weinstein, Henri Niget, Colin Arrault, Guillaume Peria, Rana Taha, Aude Azzi, Noah Kiprob Langat, Priscilla Ayeung, Yi-Fan Chen, Kristen Fitzpatrick, Abigail Sandler, Justine Hager, Hannah Stollerey, Sky Achitoff, Christine Giorgio, Ricardo Souto, Kris Li, Angela Sun, Michael Delgado, Taylor Urbshott, Caitin Sills, Michael Hoehn, Shuya Tang, Jenn Kim, Mark-Henry Jean Decrausaz, Kate McNamara.

CASE STUDY:
Research by Design: The Gantry

Hawkins\Brown

We believe that buildings are facilitators for positive change. Not only for those that live, work and play in them, but the actual process of designing and building can be a huge, positive opportunity for change for those that create them as well. At Hawkins\Brown we have been using building projects as frameworks to research new ways of designing and constructing, informing both practice and industry.

The core members of our Digital Design team each studied architecture at the Manchester School of Architecture. Throughout the master's training they were inspired to use built environment data alongside prototyped, computational tools to form architectural responses. Hawkins\Brown employed several of these graduates over successive years from 2013, which proved to be fertile ground for budding archi-programmers who began to explore these approaches on live projects.

These graduates were embedded into projects in the 'traditional' sense, operating as architectural assistants. However, with an inkling that computational design could benefit the project, they would dedicate around 80% of their time to typical architecture roles and the remaining 20% to developing computational design techniques. The first explorations of computational design began with the early stages of Here East,[1] through the 'dazzle' frit pattern and CNC-milled[2] timber lattice meeting pods at the Broadcast Centre.[3] Here East was about creating 'London's new home for making' through innovation, creativity and technology, and it was critical to embody this in how we delivered the project. Sitting at the axis between computational design and digital fabrication, The Gantry, a series of 23 mass-custom-built artist studios at Here East, exemplifies this and provided a vehicle for researching disruptive technologies that have further informed how we practise.

Hawkins\Brown, Here East, London, 2018. The Here East 'dazzle' facade, inspired by camouflage used to paint ships during the First World War, was computationally designed using Grasshopper. The design tool allowed the team to rapidly visualise the dot matrix pattern at both a super-graphic scale and at a human scale, and ensure that it met necessary solar shading criteria. Once fixed, the print instructions could be broken down by facade panel and sent directly for ceramic frit printing.

Left: Hawkins\Brown, Here East, London, 2018. The shared suspended meeting pods sit as intriguing objects within the collaborative atria spaces. Lattice geometry for each pod was computationally designed and set out using Grasshopper. The final elements were prefabricated using CNC milling and were assembled on-site.

Below: The challenge was to create a system to generate the desired variety over 23 iterations, using computational design. Each iteration would then be sent for fabrication, with parts delivered to site for assembly. This design and delivery system diagram was a fundamental project plan outlined at the very beginning of the project.

The Gantry

The vision for the project was to create a 'cabinet of curiosities' for the twenty-first century, a series of unique, playful, lightweight structures placed on to an existing, heroic steel gantry overlooking the Queen Elizabeth Olympic Park. It was a chance encounter between designers from Hawkins\Brown and Architecture 00, the incubator for WikiHouse,[4] that proved just the happenstance we needed for the perfect twenty-first-century Wunderkammer.

Hawkins\Brown and Architecture 00 collaborated to develop the WikiHouse system into a flexible parametric design tool, where a user could easily alter parameters for a building[5] and watch the 3D model regenerate live. At the time, the design system could not rapidly generate 23 bespoke forms for fabrication; each would need to be uniquely designed. Furthermore, WikiHouses had never been delivered at this scale before.

How to solve this problem? Computational design was the answer. CNC machines can motor through cutting tasks if you generate

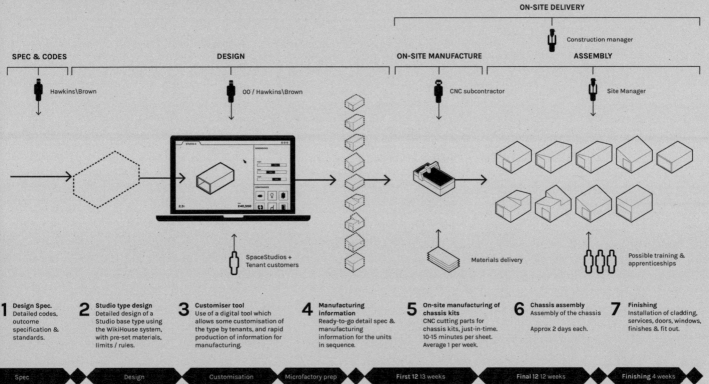

ON-SITE DELIVERY

Construction manager

SPEC & CODES	DESIGN	ON-SITE MANUFACTURE	ASSEMBLY
Hawkins\Brown	00 / Hawkins\Brown	CNC subcontractor	Site Manager

SpaceStudios + Tenant customers

Materials delivery

Possible training & apprenticeships

1 Design Spec.
Detailed codes, outcome specification & standards.

2 Studio type design
Detailed design of a Studio base type using the WikiHouse system, with pre-set materials, limits / rules.

3 Customiser tool
Use of a digital tool which allows some customisation of the type by tenants, and rapid production of information for manufacturing.

4 Manufacturing information
Ready-to-go detail spec & manufacturing information for the units in sequence.

5 On-site manufacturing of chassis kits
CNC cutting parts for chassis kits, just-in-time. 10-15 minutes per sheet. Average 1 per week.

6 Chassis assembly
Assembly of the chassis

Approx 2 days each.

7 Finishing
Installation of cladding, services, doors, windows, finishes & fit out.

| Spec | Design | Customisation | Microfactory prep | First 12 13 weeks | Final 12 12 weeks | Finishing 4 weeks |

6

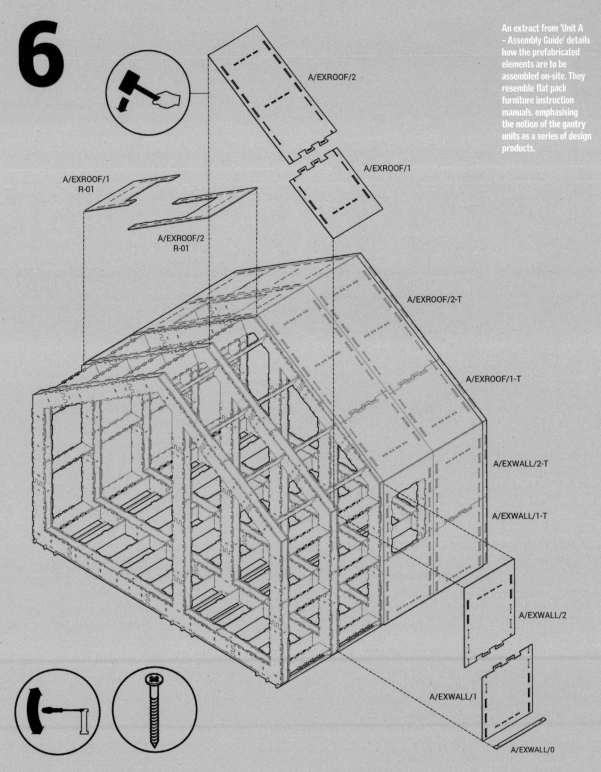

An extract from 'Unit A – Assembly Guide' details how the prefabricated elements are to be assembled on-site. They resemble flat pack furniture instruction manuals, emphasising the notion of the gantry units as a series of design products.

A/EXROOF/2

A/EXROOF/1

A/EXROOF/1
R-01

A/EXROOF/2
R-01

A/EXROOF/2-T

A/EXROOF/1-T

A/EXWALL/2-T

A/EXWALL/1-T

A/EXWALL/2

A/EXWALL/1

A/EXWALL/0

WINDOW LOCATIONS

The location of windows and rooflights shown here is only indicative.
For the specific location of sheathing panels with window or rooflight
openings (labelled W or R) for each plot refer to the architectural drawings.

The team chose to use Rhino for the fabrication model due its superior geometry engine – it could process many complex geometries more quickly than Revit – and Revit for the design model due to its built-in functionality for producing typical design information.

The design system comprised two custom-built computational models, simultaneously controlled using a single spreadsheet of editable design parameters for each structure.

Flat pack timber elements arrive on site for assembly with components and connection locations coded consistently with the Assembly Guides. Each timber portal frame is erected like a barn raising, with sheathing boards bracing the frames together and connected using crush fit joints.

all the unique cutting files needed to feed them. Therefore, it was imperative that our design model embedded every joint, detail and associated cutting instructions for each WikiHouse. Files for fabrication could then be automatically regenerated, at the push of a button, with each design change. This front loaded most of RIBA Stage 4 – Technical Design – which we completed once, rather than 23 times for the unique studios.

The design system comprised two custom-built computational models, simultaneously controlled using a single spreadsheet of editable design parameters for each structure. The first model was built in Rhino 3D, controlled using Grasshopper,[6] and the second built in Revit, with adaptive components, controlled using Dynamo.[7] The Rhino/Grasshopper model was generated at a fabrication level of detail in 3D, with all required timber components simultaneously set out on to 2D cutting sheets. The Revit/Dynamo model would be used to produce more typical architectural drawings and schedules for planning, client review and tender. Very much a working prototype, there could be many more improvements of the system, such as developing a more intuitive user interface or streamlining a direct connection from fabrication model to factory.

As designers, we were incredibly close to the construction process, providing assembly guides for each unique structure. Even the layout of the WikiHouse components on to a plywood sheet were nested programmatically for material optimisation and to expedite manual extraction of the parts in the factory. The main contractor conducted the project like a manufacturer, producing multiple iterations of a product, honing it by building physical prototypes and then, once satisfied, proceeding with full fabrication.

Feeding back

The Gantry is metaphorical in more ways than one. The finished article is a support framework in which its inhabitants create the innovative disruptions of the future, while its design and delivery established something very similar: supporting a growing computational culture in practice. Our Digital Design team was formalised off the back of it and continues to grow. Now 100% dedicated to Digital Design projects (computation and new technologies), team members divide their time into thirds between project support, practice-wide RD&I[8] and internal and external knowledge sharing. Alongside delivering increasingly sophisticated programming expertise, the team aims to create a practice-wide culture of computational thinking.

This is not without its challenges. Practices need to recruit and train computational talent in order to take advantage of the growing opportunities of Digital Design. For us this is no different – we must employ, educate and embed computational talent into our projects if we wish to conjure up the next Gantry. Immersion in projects helps stimulate and drive the most wholesome digital innovation, and we believe a framework for engagement with computation is therefore vital. As these projects develop, it is then imperative to feed the knowledge and skills gained back into practice. As each project innovation cycle progresses, it lays the foundations for others to begin. The result, we hope, will be continual and concurrent cycles of digital innovation and improvement for both projects and practice.

We speculate that future computational design will come in three doses:

1. A small dose for all designers, from Partner to Part 1, understanding emerging technologies enough to know what it can offer their projects.
2. Medium doses for computational designers, in the trenches amongst project teams, capable of initiating, developing and delivering technology-backed ideas on projects.
3. The largest doses go to the Digital Design team, supporting computational designers and dedicating their time to the most sophisticated digital initiatives.

At Hawkins\Brown, Digital Design feedback using the knowledge gained supporting past projects to help deliver future ones. In order to upskill staff, they deliver focused workshops teaching architects Computational Design in Grasshopper, automation routines in Dynamo and introductory Python. Regular CPD, presentations and shared code resources help to continually stimulate a softer, wider, cultural digital improvement.

As each project innovation cycle progresses, it lays the foundations for others to begin.

The final gantry spaces are let to creative and innovative start-up businesses. Define Engineers, specialists in complex geometry and computational design, is one such business and will be pushing emerging technologies in engineering and construction.

1. Here East is a ground-breaking scheme, designed and delivered by Hawkins\Brown, to transform the former Press and Broadcast Centre on the Queen Elizabeth Olympic Park into 1.2 million m2 of commercial space for London's creative and digital industries. The first phases of the project were completed in 2018.
2. CNC milling (computer numerical control milling) is a machining process which uses computerised controls and rotating multi-point cutting tools, progressively removing material from a workpiece to produce a custom-designed element.
3. The largest building on the Here East site comprising of 61,409m2 of flexible workspace for businesses, educators and entrepreneurs who routinely push boundaries in each of their respective industries.
4. WikiHouse is a digitally manufactured building system made up of adaptable, standardised, lightweight plywood parts, precision manufactured using CNC milling machines and capable of creating a variety of forms. https://www.wikihouse.cc/
5. Adjustable parameters include form, footprint, roof pitch, height, door and window/rooflight locations and sizes, wall thickness, internal column location and over-claddings.
6. Grasshopper is a plug-in for Rhinoceros 3D modelling software. It provides a visual interface for building algorithms that generate geometry in Rhino. https://www.grasshopper3d.com/
7. Dynamo is a plug-in for Revit BIM software. It provides a visual interface for building algorithms that generate geometry and manipulate data in Revit. https://dynamobim.org/
8. Research, Development and Implementation.

CASE STUDY:

Digital Constructivism:
Democratising the Digital

LASSA Architects

LASSA Architects, Villa Ypsilon,
Greece, 2017. South view of the
villa's concrete shell.

'*What we do depends on who we are, but it is necessary to add also that we are, to a certain extent, what we do*' – Henry Bergson, Creative Evolution The project presented here was a self-funded experiment aimed at developing new working methods that democratised the use of digital design and manufacturing techniques, as well as their deployment on ordinary commissions. The experience required a complete rethink of the role of the architect during the design, procurement and assembly stages while enhancing creativity. The project demonstrates the viability of non-standard construction using both digital design and manufacturing within economically tight constraints. It provided an opportunity to meet complex project demands – in terms of form, structure and budget constraints (approximately €1500/m2) – while concurrently redesigning the workflow and communication of complex details to suit the remote location and local workmanship.

The project demonstrates the viability of non-standard construction using both digital design and manufacturing within economically tight constraints.

Project Description

The Villa Ypsilon is nested within the topography of an olive grove in Greece. It is composed of a concrete shell roof that acts as an accessible extension of the terrain and frames specific views of the site from inside and out. The roof's bifurcating pathways define three courtyards that form distinct hemispheres with specific occupancy depending on the course of the sun. The height of the house is limited to the top of the olive trees to enable integration with the surrounding landscape. The circulation through, around and on top of the house forms a continuous promenade comprising indoor and outdoor activities. The iterative design of the shell through its shadow analysis was aimed at designing occupancy and activating the use of the courtyards throughout the day. The resulting environmental strategy of the project favoured the development of climate-resilient geometry rather than the use of mechanical systems or industrial products from a catalogue. The form of the concrete shell, coupled with the planted roof and cross-ventilation strategy, balances the reach of sunlight into the house across the summer and winter seasons and maintains a natural level of comfort.

Enabling complexity while controlling quality and cost

Very early on, a combination of remote location, limited budget and the absence of skilled labour induced a construction strategy that called for a large amount of off-site prefabrication. The main construction problem was the design and production of the formwork for the casting of the concrete shell. The acquisition and use of an in-house CNC machine allowed for the production of all non-standard elements. This included foundation guides, concrete shell formwork, living room lost formwork and acoustic ceiling, west facade custom Corian window frames, interior furniture and partition systems, milled marble showers, as well as landscape and pool formers. A 'hands-on' approach allowed for greater control over costs while minimising the use of commercial off-the-shelf industrial

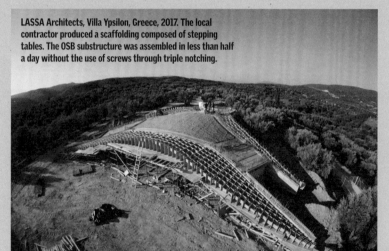

LASSA Architects, Villa Ypsilon, Greece, 2017. The local contractor produced a scaffolding composed of stepping tables. The OSB substructure was assembled in less than half a day without the use of screws through triple notching.

LASSA Architects, Villa Ypsilon, Greece, 2017. View of the finished formwork. The top layer of the formwork was composed of plywood for the exposed concrete overhangs.

products, favouring instead a local supply chain for the sourcing of concrete, terrazzo and marble.

The formwork was entirely developed in LASSA's office. It was composed of 5,200 different parts that were cut from 500 panels of OSB or plywood of various thicknesses. The presence of the machine in the office enabled extensive prototyping of all parts of the building as well as a 1:1 load testing for the main shell formwork. These were then digitally cut and labelled in house at the architect's practice.

Perhaps the most innovative part of the project was the development of the double curved lost formwork. This element was designed to take double curvature through the use of 6mm plywood on the top. The thin wood shell was formed and supported by 18mm arching plywood fins which also act as an acoustic ceiling after the demoulding; the fins had protruding dowels that were used to attach the lost formwork to the rebar. The integration of lights in the ceiling required the production

LASSA Architects, Villa Ypsilon, Greece, 2017. The architects produced a 1:1 prototype to understand the logistics involved and the assembly time. A load test of up to 1.5 tons was also performed to better understand the structural stability of the formwork. Through a collaboration with our structural engineer, Manja van de Worp, we were able to reduce the size of the members to economise wood.

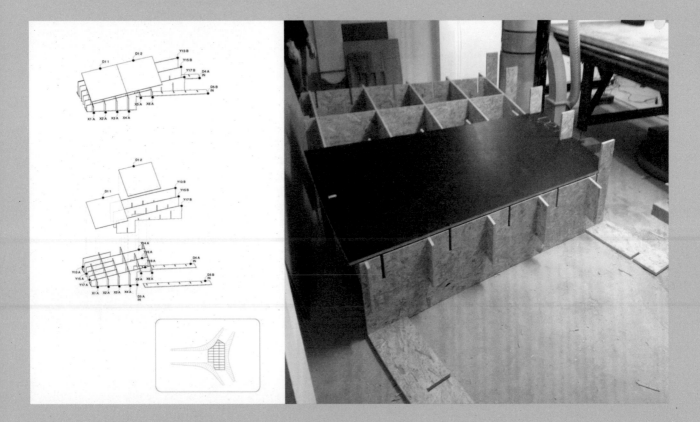

of CNC milled moulds, which were used to thermoform the bespoke roof light diffusers.

Partial self-assembly

The parts were shipped to the site and were partially self-assembled by the architects themselves, with a team of four, in just four days. The formwork was designed in such a way that its assembly did not require the use of screws or tools. Particular attention was paid to the development of an intuitive labelling system that would prevent the need for an instruction manual, which would significantly reduce assembly time. The construction of the project was effectively paperless and did not require the use of plans. Instead of using plans, sections and elevations as building instructions, the architects produced wood formers and formwork directly in the office. This method allowed a reduction in construction time to seven months without compromising quality or exceeding the budget.

The formwork was designed in such a way that its assembly did not require the use of screws or tools.

LASSA Architects, Villa Ypsilon, Greece, 2017. Upside-down view of the lost formwork prototype produced in the architect's office. The top layer is composed of 6mm plywood and the ribbing of 18mm ply

Opposite: LASSA Architects, Villa Ypsilon, Greece, 2017. Installation of the acoustic ribbing substructure which was also used to form the curvature of the top wood shell.

Below: LASSA Architects, Villa Ypsilon, Greece, 2017. 6mm plywood panels were notched on top of the 18mm plywood reading. The protruding walls are used to attach the last formwork to the rebar.

Bottom: LASSA Architects, Villa Ypsilon, Greece, 2017. Interior view of the lost formwork. The practice's approach to design integration is exemplified here by the incorporation of the forming function, lighting and acoustics. The curving seams provided an economy of wood, as smaller ribs helped improve the nesting.

Minimising the use of off-the-shelf building products in favour of locally sourced materials can help to form the beginning of a response to systemic environmental and social problems.

Redesigning the architect's role: new workflows within construction practices

The integration of product libraries within our digital tools challenges known forms of practice in an environment that seeks to make the specification of building systems as effortless as possible. Another consequence of automation through BIM is the standardisation of our built environment. Concurrently, the direct connection between design and manufacturing tools and their democratisation is allowing for non-standard architecture practices to exist while being financially viable.

Villa Ypsilon is part of broader research through practice conducted over four built projects using a similar methodology. LASSA's involvement in the production of these projects helped increase formal complexity and reduce the construction costs for the owner while enhancing significantly creative possibilities. Minimising the use of off-the-shelf building products in favour of locally sourced materials can help to form the beginning of a response to systemic environmental and social problems caused by industrialisation and its reliance on a transnational supply chain and opaque labour conditions. Finally, the project involved the development of the practice's business model, expanding the role of the architect in areas including production design, building parts procurement and assembly.

CASE STUDY:

Panels, Polygons and Pixels:
How Data Informs Supertall
Tower Design

SHoP Architects

SHoP Architects, 9 DeKalb, 2016. Rendering.

The hexagonal form incorporates materials and shapes derived from the 1908 Dime Savings Bank, the landmarked building from which it rises.

The process of bringing an architectural project from design concept to built environment is long and complex. SHoP Architects started work in 2014 on 9 DeKalb, a supertall tower in downtown Brooklyn, scheduled for completion in late 2021. This life cycle offers endless possibilities for experimentation and innovation, methods and ways forward. The key for progression is improving on this process and the evolution and application of technology in the architecture, engineering and construction (AEC) community. While industries like aerospace and automotive have invested in, and calibrated, the design-to-fabrication cycle over generations, the AEC industry has not often been an early adopter; nor has it looked to other sectors in contributing to a better built environment.

Tremendous insight can be gained from these other industries. Their advances and successes can inform strategies for creating and realising architecture, as we have done throughout our work with 9 DeKalb. New technologies allow for innovation across the wide spectrum of the design to construction process. In the case of 9 DeKalb, these technologies span and connect two centuries: the hexagonal form incorporates materials and shapes derived from the 1908 Dime Savings Bank, the landmarked building from which it rises, on a narrow, shared site. Vertical extrusions of black, gold, copper stainless, anodised steel and glass panels exchange size and prominence, creating a gradient effect to the 327-m top of the first supertall tower in downtown Brooklyn.

Finance to Form

Throughout the design process, our team – a multidisciplinary group of architects, engineers and software experts – resolved

SHoP Architects, 9 DeKalb, 2016. Rendering.

and customised data management and interaction for the architectural process in several different forms. In the early stages, the most important parameters were defined by the 9 DeKalb residential unit mix. As the ownership revised the proforma to reflect market research and confidence in the project, our design model needed to be accordingly adaptive in order to recalibrate the mix of condos, market rate and affordable rental units.

The spreadsheet of unit types and areas provided a tool to communicate between financial parameters and architectural constraints. We created a series of Grasshopper scripts to translate the numerical data into a 3D massing and then populate that massing with the latest iteration of the envelope, structural layout and other cost vectors. Advancing these architectural expressions in metadata and in 3D early in the process allowed the team, client and other stakeholders to make collaborative visual and quantitative decisions on the project in real time.

SHoP Architects, 9 DeKalb, 2016. Facade optimisation analysis.

Parallel Envelope Models

As the project progressed through documentation, we maintained several parallel models to manage the complexity and speed of the project. This strategy is particularly vital to the tower facade, a unitised curtain wall of glass and opaque panels that references the spirit of Art Deco skyscrapers and the 1908 bank building. Each of the 4,000 opaque panels is composed of a series of glass and steels (black, gold, copper stainless and anodised). Our challenge for the facade was to preserve the design intent of a smooth gradient while reducing the number of unique components down to 85 panel types of varying width and extrusion composition.

During documentation, we maintained the facade in multiple forms: a material-mapped rendering model (Rhino), a detailed information model (Revit) and a quantified-data model (Filemaker). To facilitate various studies and outputs, we developed automated processes to pass lightweight information between each model. For instance, the panel count and resulting gradient after

Unit Configurations

System Efficiency

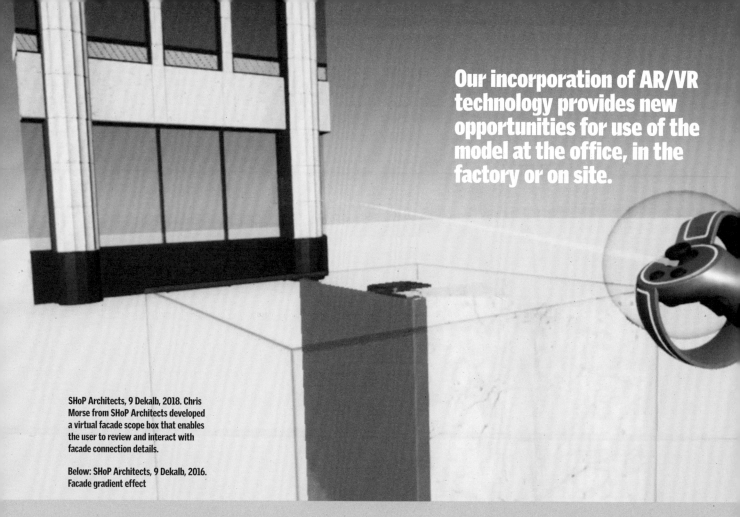

Our incorporation of AR/VR technology provides new opportunities for use of the model at the office, in the factory or on site.

SHoP Architects, 9 Dekalb, 2018. Chris Morse from SHoP Architects developed a virtual facade scope box that enables the user to review and interact with facade connection details.

Below: SHoP Architects, 9 Dekalb, 2016. Facade gradient effect

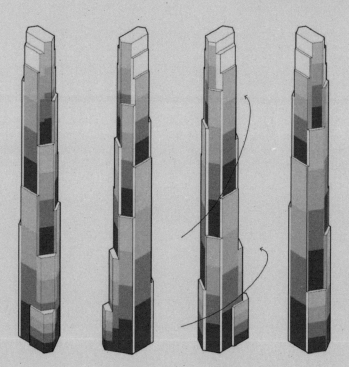

several iterations was coordinated to meet the contractor's production, storage and installation schedules. The processes we employed allowed visual effects, technical constraints and construction impacts to be studied simultaneously.

The Data-Rich BIM Model

Our development of a data-rich Building Information Model (BIM), built to meet many demands during the design process, became increasingly valuable as the project moved into construction. While the at-desk benefits of BIM are already widely established, our incorporation of AR/VR technology provides new opportunities for use of the model at the office, in the factory or on site. The 9 DeKalb rendering model, federated documentation model and curtain wall panel database provided the source material for several case studies in augmented and virtual reality (AR/VR) implementation.

SHoP Architects, 9 Dekalb, 2018. The team created an augmented reality application that allows the user to drop the building into location at a construction site for viewing within its future context.

Below: SHoP Architects, 9 Dekalb, 2019. A SHoP Architects' app allows overlay of a digital twin on top of a physical mock-up to verify connection details and design intent.

Increased Communication via Custom Applications

The ultimate priority of SHoP Architects' innovations in AR/VR software development is to facilitate a level of real-time communication and collaboration across design teams, consultants and stakeholders – the communication that can determine the success of an entire project. Solving problems before they arise pre-empts costly and time-consuming change orders. SHoP Architects' 9 DeKalb design team worked closely with the software development team to build a range of prototype tools to optimise communication. These tools use different media and platforms depending on the problem at hand, including AR/VR, mobile development and the web.

The iterative nature of this project presented many design options for facade materials and connection details. These were much easier to consider with the assistance of an immersive and interactive VR facade review tool. This innovative visualisation gave the team the ability to analyse and, when now evident, rethink the detail and finalise the design. After SHoP Architects built the physical mock-up, AR allowed the team to view it within the context of the entire facade, and overlay and examine the connection details, as well as experience the building in place within the larger context of the site and the city.

After construction started on site, SHoP Architects began working on a tool for quality assurance and information querying. We built this to give teams on the ground the information they needed quickly, making it possible via immediate access to the embedded BIM data and AR. This was in huge contrast to the traditional on-site task of flipping pages full of drawings or looking at a large PDF. As construction continued, the COVID-19 pandemic made physical group site visits challenging. In response, SHoP Architects built a web application that allows teams to quickly scan areas of the construction site and distribute those scans to their colleagues, partners and other stakeholders. The web AR technology even allows a user to drop these same scans on to, say, the floor of their living room at home, giving remote team members the ability to inspect work and progress in far more depth than photos and videos would offer.

SHoP Architects built a web application that allows teams to quickly scan areas of the construction site and distribute those scans to their colleagues, partners and other stakeholders.

SHoP Architects, 9 Dekalb, 2020. Tim Li from SHoP Architects helped to develop the augmented reality software. This application leverages BIM data to give users necessary information in real time via a document query system to increase efficiency and collaboration on and off site.

Final Word

Kostas Terzidis

Architecture is a peculiar area of design, as it stands between idea and reality. It bridges the gap between the abstract concepts of aesthetics, harmony and imagination with the rigid facts of materiality, construction and gravity. Traditionally, the process of transforming ideas into reality was guided by knowledge and intelligence from talented people. This well-established process was disrupted about 50 years ago by a machine called the computer. While not human, it demonstrated a capability to address and solve quantitative tasks in speeds and complexities far superior to that of humans.[1, 2]

For most designers, computers are regarded as tools that aid the design process. However, computers as tools can affect the intellectual process of design. It is a powerful tool with arithmetic and logical capabilities. As such, it can lead to a lack of originality in the process of design by CAD tools.[3] The problem is that often designers have not sought to create, adapt and control these tools, instead becoming mere users of the tools and followers of the technology.

With the advent of artificial intelligence in the last five years, computers evolved from tools into entities capable of demonstrating intelligent behaviour. Many white-collar jobs were replaced by new technology and concern spread about the fate of the designer. What if the designer gets replaced one day too? Graduates from schools of design have become less and less cognisant of the emerging technologies, while computer science school graduates have become more and more adept, increasingly controlling indirectly the field of design by taking over CAD, VR, AR, robotics, gaming and simulation, and eventually monopolising all design tools.

Design needs intelligence to manifest itself and intelligence needs design to apply itself. Both are converging in such a way that eventually one day they may meet.[4] Meanwhile, the convergence leaves fewer and fewer tasks for the designer, while automating the rest. This is seen by many as a danger to the theoretical, ethical, economic, social or even political dimension of design. It is a crisis of design that many see as increasingly inevitable, as an upcoming 'tsunami'.[5] So, then what to do? What to advise the new students or the young faculty to avoid the coming crisis?

The first step to study any word is to define it. Let us start with the definition of 'design'. In Greek, the etymology of the word 'design' (σχέδιο) means 'to have lost' (έσχειν).[6] According to this definition, design is not about making new things that appear to come from the designer's mind but to remember things that already existed in the mind but were lost.[7]

The second step is to define what we mean by 'mind'. Again, in Greek the word 'mind' (νους) derives from the meaning 'to arrange' (νέμω).[8] According to this definition, the mind is a vast arrangement mechanism of information that pre-exists, some of which is hidden and needs to be revealed. Through the logic of language and the techniques of mathematics, we can extract, discover and organise the information previously hidden.[9] When a mind is successful in solving a problem, we refer to it as intelligent or smart. But does that occur only in the human mind or perhaps elsewhere? For example, in a machine?

The third step is to define what we mean by a 'machine'. The etymology of the word 'machine' (μηχανή) means 'therapy' (μάω+άκη) in Greek.[10] According to this definition, a machine is a device to cure a problem. So, if the problem of design is about forgetting (according to our earlier definition), to be successful a machine should be able to organise, search and reconstitute memory arrangements.

A

B

C

D

With the advent of artificial intelligence in the last five years, computers evolved from tools into entities capable of demonstrating intelligent behaviour.

All the possible arrangements of a simple four-fixture bathroom are 65,536. The non-repetitive, rotationally specific arrangements are 384 (a). However, after eliminating all arrangements that have a toilet seat facing a door (b) and eliminating any arrangement that uses more than 6m of pipeline, the number of successful bathrooms is only 12 (c). Detail of succesful bathroom arrangement (d).

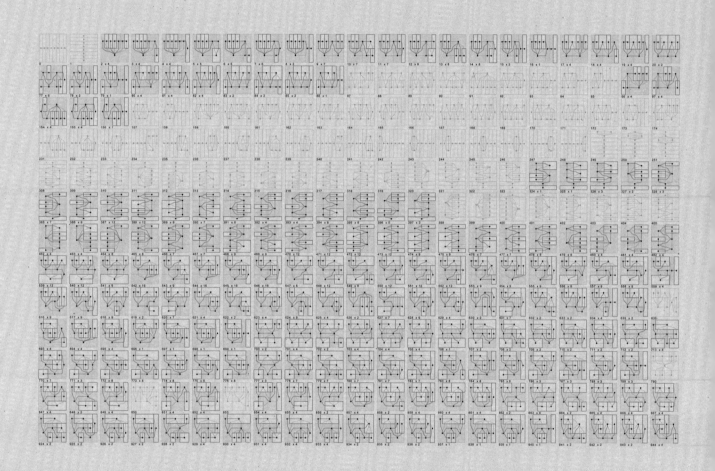

An architectural structure can be expressed
by topological interrelationships between
spaces. Such a system has the capability of
generating a set of all possible permutations
that oscillate between order and chaos. For
instance, 26,8435,456 patterns of possible
permutations can be generated from eight
units of space, as seen partially here.

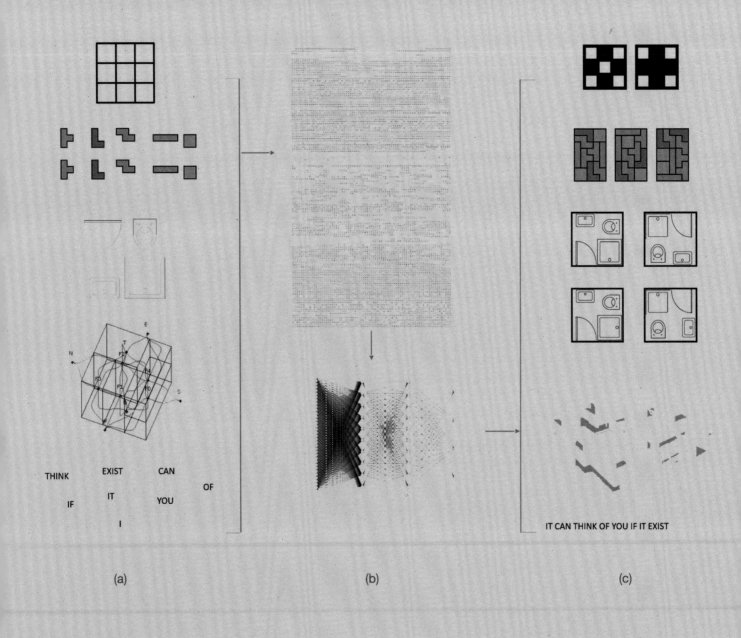

THINK   EXIST   CAN

IF   IT   YOU   OF

I

IT CAN THINK OF YOU IF IT EXIST

(a)   (b)   (c)

**Deep Permutation Design scheme.
Elements (a) combined into permutations
and extracted through a trained neural
network (b) to select the best case (c).**

Designers need to be the change rather than following the change. They must switch from being tool users to become toolmakers.

In English, the words 'design', 'extract' or 'machine' are both nouns and verbs. If seen as nouns, then they do not necessarily need human intention. They are just things. Things have no intention. So, in that sense, machines without intention can theoretically behave in a smart manner and therefore can design. But how?

The key to this is to explore the word 'arrangements'. If the human mind is a vast mechanism of arranging information, then we need to investigate how these arrangements occur, what they mean and how many are they. This is where computers are useful: they can compute all possible arrangements, revealing both known and unknown arrangements and search for the best ones using AI.

The premise is that design is not about making something out of nothing, but rather about extracting something out of everything.[11] AI tools can be applied to any design-related field, including architecture, music, fashion, cooking, story-telling, gaming, typography, art and industrial/service design.

Returning to the original question: what should designers do about the upcoming AI tsunami?

The answer is simple but hard. As in a real tsunami, the safest place to be is on top of the wave. That is the only safe place. Heraclitus once said that 'the only constant is change' (μεταβάλλον αναπαύεται).[12] Designers need to be the change rather than following the change. They must switch from being tool users to become toolmakers. Tools are powerful, complementary and controlling vehicles that allow designers to explore unchartered territories of design. Toolmaking is the only way for designers to take control of their fate into their own hands.

1 Negroponte, N., *The Architecture Machine*, MIT Press, Cambridge, Mass, 1970.
2 Mitchell, W., *Computer-Aided Architectural Design*, Petrocelli/Charter, New York, 1977.
3 Serraino, P., 'Form Follows Software', *Arcadia Proceedings*, vol. 22, 2003 pp. 186–205.
4 Kurzweil, R., *The Singularity is Near: When Humans Transcend Biology*, Viking, New York, 2005.
5 Kelsey, T., *Surfing the Tsunami*, Todd Kelsey, Illinois, 2018.
6 Liddell, H., *An Intermediate Greek-English Lexicon: Founded upon the Seventh Edition of Liddell and Scott's Greek-English Lexicon*, Benediction Classics, Oxford, 2010.
7 Terzidis, K., 2006. *Algorithmic Architecture*, Architectural Press, London, 2006.
8 Liddell, H., *An Intermediate Greek-English Lexicon: Founded upon the Seventh Edition of Liddell and Scott's Greek-English Lexicon*, Benediction Classics, Oxford, 2010, p. 1607.
9 Plato, *The Dialogues of Plato*, Thoemmes Press, Bristol, 1997.
10 Liddell, H., *An Intermediate Greek-English Lexicon: Founded upon the Seventh Edition of Liddell and Scott's Greek-English Lexicon*, Benediction Classics, Oxford, 2010, p. 1537.
11 Terzidis, K., *Permutation Design: Buildings, Texts, and Contexts*, Taylor & Francis, London, 2014.
12 Kahn, C., *The Art and Thought of Heraclitus: An Edition of the Fragments with Translation and Commentary*, Cambridge University Press, Cambridge and New York, 1979.

Contributors

May Bassanino has over 15 years' experience in academia, conducting and delivering interdisciplinary projects in the areas of collaborative engineering, process modelling and evaluation frameworks. May joined the [CPU] lab at the Manchester School of Architecture as a project manager and researcher in 2019 to work on urban complexity research projects.

David Benjamin is Founding Principal of **The Living**, Director at Autodesk Research and Associate Professor at Columbia GSAPP. He and the firm have won many design prizes, including the Emerging Voices Award from the Architectural League, the Young Architects Program Award from the Museum of Modern Art, and a LafargeHolcim Sustainability Award.

Adam Chernick is the Director of Interactive Visualization at **SHoP Architects**. His experience on the design technology and documentation side of AEC helps to guide software development efforts at SHoP. He has designed, built and launched apps for both iOS and Android, and is a frequent speaker on the implications of emerging technologies within AEC. His applications have been featured in publications including the *New York Times*.

Eric Cheung is a UK-qualified architect and researcher at the Manchester School of Architecture, in the [CPU]lab. He has taught architecture at the University of Nottingham and Manchester School of Architecture. He specialises in computational methods and custom tools to understand dynamic urban systems and spatial design. His research explores urban land-transport integrated systems related to spatial design and planning.

Angelos Chronis is the head of the **City Intelligence Lab** at the Austrian Institute of Technology in Vienna and teaches at the Institute for Advanced Architecture of Catalonia and the Bauhaus University in Weimar. His research focuses on performance-driven design and embedding computational intelligence into design systems.

Mollie Claypool is an architecture theorist working on issues of social justice, concerned with the implications of new technologies and automation on architectural production and disciplinary social practices. She is Co-Director of **AUAR Labs** at The Bartlett School of Architecture, UCL and Director of Automated Architecure Ltd (AUAR), a design and technology consultancy.

Theodore Galanos is a senior researcher at the **City Intelligence Lab** at the Austrian Institute of Technology in Vienna and a leader in computational technologies in design for the built environment. He works at the intersections of design, data, and intelligence, and seeks to create innovative, data-driven solutions within the field of Computational Environmental Design.

Soomeen Hahm is the founder of the **SoomeenHahm Design Ltd**, a design faculty and robotic researcher at the Southern California Institute of Architecture (SCI-Arc). Interested in looking at the ecology of computational power, technology and human intuition, the office has been focusing on the use of augmentation (i.e. AR/VR/wearables) to invoke immersive experiences of users and builders of our future cities.

Hawkins\Brown – **Jack Stewart, Ben Robinson** and **Ben Porter**. Digital Design at Hawkins\Brown is a team of architects turned software developers that offer expertise across design projects. The team use focused scripting and coding expertise to help resolve design challenges and create tools to deliver efficiently more robust production information.

Theo Sarantoglou Lalis is co-founder of **LASSA Architects**, an international architecture practice with offices in London and Brussels. He has been teaching a diploma unit at the Architectural Association in London since 2009. He has taught postgraduate studios at Harvard and Columbia and undergraduate studios at Chalmers and LTU in Sweden.

Phil Langley is an architect and computational designer based in London and leads the Creative Techonoliges team at **Bryden Wood**. Bryden Wood is a leading practice in Design for Manufacture and Assembly (DfMA) and the Creative Technologies team specialises in design automation. It combines DfMA approaches and cutting-edge algorithmic methods to build digital technology that accelerates the delivery of social and economic infrastructure.

Danil Nagy is a designer, educator and entrepreneur creating technology to transform the building industries. He teaches architecture and technology at Columbia University and Pratt Institute, and founded Colidescope, a consultancy focused on bringing automation and digital transformation to the AEC industry. As CTO of iBuilt, he oversees the development of ground-breaking technologies to change the way buildings are designed, built and managed.

Stefana Parascho is a researcher, architect and educator whose work lies at the intersection of architecture, digital fabrication and computational design. She is currently an assistant professor at Princeton University, where she founded the CREATE Laboratory Princeton and is co-leading the PhD programme in Technology at Princeton's School of Architecture.

Gregg Pasquarelli is a founding principal of **SHoP Architects**, where he has led some of the firm's most complex and influential projects, including The Porter House, Barclays Center, the East River Waterfront Esplanade and Pier 17, the American Copper Buildings and the second-tallest tower in Manhattan, 111 West 57th Street.

Jens Pedersen is a computational architect who is currently conducting an industrial PhD, which is a collaboration between **Odico Construction Robotics** and the Aarhus School of Architecture. The PhD explores how Odico's Factory on the Fly framework can be augmented with timber-processing capabilities and become part of the growing timber construction sector.

Craig Rosman is an architect in New York City. He received an M.Arch II from Yale University and a B.Arch from Carnegie Mellon University in Pittsburgh, PA. He is currently a senior associate at **SHoP Architects**, working on a variety of projects with a focus on applying digital design techniques to large urban mixed-use developments.

Ulysses Sengupta is a reader at the Manchester School of Architecture. He is the founding director of the Complexity Planning and Urbanism research laboratory **[CPU]lab** and co-leads the undergraduate and master's design atelier **[CPU]ai**. He uses a complexity framework for transdisciplinary research spanning natural, social and design sciences. This involves new digital tools, computational thinking and theory addressing urban transformations.

Foteinos Soulos is a senior associate at **SHoP Architects**, currently managing 9 Dekalb. He specialises on facade systems and has been the facade coordinator through several projects at SHoP, such as Pier 17, the US Embassy in Honduras and other government projects around

the world. He is also involved in various app development efforts using virtual and augmented reality technologies.

Solon Solomou is an associate researcher at the Manchester School of Architecture, in the **[CPU]lab** and teaches in the master's design atelier **[CPU]ai**. His PhD research involves the creation of digital tools to understand and model the development of cities. Using an ABM methodology, complexity theories and data, he explores complex urban phenomena towards effective design intervention.

Asbjørn Søndergaard is founding partner and Chief Technology Officer in **Odico Construction Robotics**, a technology enterprise dedicated to large-scale architectural robotic formwork fabrication. Founded in 2012, in a joining of research trajectories following the Fabricate 2011 conference, Odico Construction Robotics has embarked on a mission to revolutionise global construction.

Jim Stoddart is a designer at **The Living**, an Autodesk Studio and a senior research scientist in the AEC Industry Futures group within Autodesk Research. His work focuses on applications of novel technologies to real-world design problems, including new materials prototypes, generative design, experimental fabrication and spatial data visualisation.

Mahmud Tantoush is a researcher at the Manchester School of Architecture, in the **[CPU]lab** and architecture studio tutor. His interests lie in computational design, smart cities, big data, geographic data science, urban morphology and complexity theories.

Kostas Terzidis is a professor at D&I at Tongji University in Shanghai and director of the **ShangXiang Lab**. He was previously an associate professor at Harvard's GSD. He is author of four books: *Permutation Design* (Routledge, 2014), *Algorithms for Visual Design* (Wiley, 2009), *Algorithmic Architecture* (Architectural Press, 2006) and *Expressive Form* (Spon, 2003).

Mette Ramsgaard Thomsen founded the Centre for IT and Architecture research group (CITA) at the Royal Danish Academy of Fine Arts, School of Architecture, Design and Conservation in 2005, where she has piloted a special research focus on the new digital-material relations that digital technologies bring forth.

Lorenzo Villaggi is a senior research scientist and designer at **The Living**, an Autodesk Research studio and an adjunct assistant professor at Columbia GSAPP. Lorenzo's work and teaching focus on novel data-driven workflows, generative systems, data visualisation and biomaterials.

Sandra Youkhana and Luke Caspar Pearson are directors of the architectural design studio **You+Pea**. They also lead the Videogame Urbanism studio at the Bartlett School of Architecture, where they promote the use of game technologies in architectural education. They have written and lectured on the subject worldwide and work with clients from the cultural, technological and games sectors on interdisciplinary design projects, consultancy and research.

Sigita Zigure is a researcher at the Manchester School of Architecture, in the **[CPU]lab**. She is interested in socio-technical processes with a particular focus on future mobility technology in cities. Currently, she is a PhD student and a research assistant on the EPSRC MaaS Prototype for TfGM Project.

Recommended Reading and Sources

Books

Alexander, Christopher, *A Pattern Language: Towns, Buildings, Construction*, Oxford University Press, 1978.

Andia, Alfredo and Thomas Spiegelhalter, *Post-parametric Automation in Design and Construction*, Artech House, 2014.

Benjamin, David et al., *Now We See Now: Architecture and Research by The Living*, The Monacelli Press, 2018.

Beorkrem, Christopher, *Material Strategies in Digital Fabrication*, Routledge, 2012.

Bernstein, Phillip, *Architecture, Design, Data: Practice Competency in the Era of Computation*, Birkhauser, 2018.

Bostrom, Nick, *Superintelligence: Paths, Dangers, Strategies*, Oxford University Press, 2016.

Bottazi, Roberto, *Digital Architecture Beyond Computers: Fragments of a Cultural History of Computational Design*, Bloomsbury, 2018.

Braidotti, Rosi, 'How To Do Posthuman Thinking', *Posthuman Knowledge,* Polity, 2019.

Brynjolfsson, Erik, *The Second Machine Age: Work, Progress, and Prosperity in a Time of Brilliant Technologies*, W.W. Norton & Company, 2014.

Burry, Jane and Mark Burry, *The New Mathematics of Architecture*, Thames & Hudson, 2010.

Carpo, Mario, *The Alphabet and the Algorithm*, MIT Press, 2011.

Carpo, Mario, *The Second Digital Turn: Design Beyond Intelligence*, The MIT Press, 2017.

Cogdell, Christina, *Toward a Living Architecture? Complexism and Biology in Generative Design*, University of Minnesota Press, 2019.

Darwin, Charles, *The Origin of Species*, Vintage, 2019.

Daugherty, Paul R. and H. James Wilson, *Human + Machine: Reimagining Work in the Age of AI*, Harvard Business Review Press, 2018.

De Angelis, Massimo, 'Grounding social revolution: elements for a systems theory of commoning', *Perspectives on Commoning: Autonomist Principles and Practices*, 2017.

Deamer, Peggy (ed.), *The Architect as Worker: Immaterial Labour, the Creative Class and the Politics of Design*, Bloomsbury, 2015.

Debney, Peter, *Computational Engineering*, The Institution of Structural Engineers, 2020.

Deutsch, Randy, *Superusers: Design Technology Specialists and the Future of Practice*, Routledge, 2019.

Flake, Gary William, *The Computational Beauty of Nature: Computer Explorations of Fractals, Chaos, Complex Systems, and Adaptation*, Bradford Books, 2000.

Ford, Martin, *The Rise of the Robots: Technology and the Threat of Mass Unemployment*, Oneworld Publications, 2016.

Foster, David, *Generative Deep Learning: Teaching Machines to Paint, Write, Compose and Play*, O'Reilly Media, 2019.

Frazer, John, *An Evolutionary Architecture*, Architectural Association, 2007.

Gershenfeld, Neil, 'How to Make Almost Anything: The Digital Fabrication Revolution', *Foreign Affairs*, November/December 2012.

Gleick, James, *Chaos: Making a New Science*, Penguin Books, 2008.

Gramazio, Fabio, Matthias Kohler and Jan Willmann, *The Robotic Touch: How Robots Change Architecture*, Park Books, 2014.

Hensel, Michael (ed.) with Achim Menges and Christopher Hight (co-eds), *Space Reader: Heterogeneous Space in Architecture*, Wiley & Sons, 2009.

Institute of Architecture, Studio Zaha Hadid, Patrik Schumacher and Zaha M. Hadid, *Total Fluidity: Studio Zaha Hadid Projects 2000–2010*, Springer Vienna Architecture, 2011.

Iossifova, Deljana, Christopher N.H. Doll and Alexandros Gasparatos (eds), *Defining the Urban: Interdisciplinary and Professional Perspectives*, Routledge, 2017.

Kaku, Michio, *The Future of the Mind: The Scientific Quest to Understand, Enhance, and Empower*, Doubleday, 2014.

Kelly, Kevin, *Out of Control: The New Biology of Machines, Social Systems, and the Economic World*, Basic Books, 1995.

King, Brett, Andy Lark, Alex Lightman and J.P. Rangaswami, *Augmented: Life in The Smart Lane*, Marshall Cavendish International (Asia), 2016.

Kolarevic, Branko and Ali Malkawi (eds), *Architecture in the Digital Age: Design and Manufacturing and Performative Architecture: Beyond Instrumentality*, 2013.

Leach, Neil and Philip F. Yuan, *Computational Design*, Tongji University Press Co., 2018.

Lipson, Hod and Melba Kurman, *Fabricated: The New World of 3D Printing*, Wiley, 2013.

Lynn, Greg, *Animate Form*, Princeton Architectural Press, 1999.

Meadows, Donella, *Thinking in Systems: A Primer,* Chelsea Green Publishing, 2008.

Melendez, Frank, Nancy Diniz and Marcella Del Signore (eds), *Data, Matter, Design Strategies in Computational Design*, Routledge, 2007.

Menges, Achim, *Material Computation: Higher Integration in Morphogenetic Design*, Academy Press, 2012.

Menges, Achim and Sean Ahlquist, *Computational Design Thinking*, Wiley, 2011.

Mitchell, Melanie, *Artificial Intelligence: A Guide for Thinking Humans*, Picador, 2020.

Negroponte, Nicholas, *The Architecture Machine: Toward a More Human Environment*, MIT Press, 1972.

Nitsche, Michael, *Video Game Spaces*, MIT Press, 2008.

Pawlyn, Michael, *Biomimicry in Architecture*, RIBA Publishing, 2016.

Pottmann, Helmut, Andreas Asper, Michael Hofer, Axel Kilian and Daril Bentley, *Architectural Geometry*, Bentley Institute Press, 2007.

Ramsgaard Thomsen, Mette, *CITA Complex Modelling*, Riverside Architectural Press, 2021.

Reiser, Jesse, *The Atlas of Novel Tectonics*, Princeton Architectural Press 1, 2006.

Roche, François, *Bioreboot: The Architecture of R&sie{n}*, Princeton Architectural Press, 2010.

Sabin, Jenny E. and Peter Lloyd Jones, Bioreboot: The Architecture of R&sie{n}, *Design Research Between Architecture and Biology,* Routledge, 2017.

Sengupta, Ulysses, Ward S. Rauws and Gert de Roo, 'Planning and Complexity: Engaging with temporal dynamics, uncertainty and complex adaptive systems', *Environment & Planning B, 43(6)*, Sage, 2016.

Sheil, Bob, Mette Ramsgaard Thomsen, Martin Tamke and Sean Hanna (eds), *Design Transactions – Rethinking Information Modelling for a New Material Age*, UCL Press, 2020.

Shiffman, Daniel, *The Nature of Code*, 2012.

Spuybroek, Lars, *Research & Design: The Architecture of Variation*, Thames Hudson, 2009.

Srnicek, Nick, *Platform Capitalism*, Polity, 2017.

Steenson, Molly Wright, 'Architects, Anti-Architects, and Architecting', *Architectural Intelligence: How Designers and Architects Created the Digital Landscape*, MIT Press, 2017.

Susskind, Richard and Daniel Susskind, *The Future of the Professions: How Technology will Transform the Work of Human Experts*, Oxford University Press, 2017.

Terzidis, Kostas, *Algorithmic Architecture*, Architectural Press, 2006.

Thompson, D'Arcy Wentworth, *On Growth and Form: The Complete Revised Edition*, Dover Publications, 1992.

Weinstock, Michael, *The Architecture of Emergence: the Evolution of Form in Nature and Civilisation*, Open University, 2016.

Weisberg, Michael, *Simulation and Similarity: Using Models to Understand the World*, Oxford University Press, 2015.

Wiener, Norbert, Doug Hill et al., *Cybernetics, or, Control and Communication in the Animal and the Machine*, The MIT Press, 2019.

Wolfram, Stephen, *A New Kind of Science*, Wolfram Media, 2002.

Woodbury, Robert Francis, *Elements of Parametric Design*, Routledge, 2010.

Yamu, Claude, Alenka Poplin, Oswald Devisch and Gert de Roo (eds), *The Virtual and the Real in Planning and Urban Design: Perspectives, Practices and Applications*, Routledge, 2017.

Yuan, Philip F., Neil Leach and Achim Menges, *Digital Fabrication*, Tongji University Press Co., Ltd, 2018.

Organisations and initiatives

Smart Geometry www.smartgeometry.org

Acadia acadia.org

Simaud simaud.org

Digital Futures www.digitalfutures.world

AEC Tech www.aectech.us

RobArch www.robotsinarchitecture.org

Index

Image Credits